Challenges and Opportunities in Medical and Apparel Textiles

Challenges and Opportunities in Medical and Apparel Textiles

Editors

Dr. Namit Gupta
Dr. V.R. Sampath
Prof. Tapan Kumar Sinha
Dr. Shamayita Patra
Dr. Syamal Maiti

Organized by

SHRI VAISHNAV INSTITUTE OF TEXTILE TECHNOLOGY
Shri Vaishnav Vidyapeeth Vishwavidyalaya, Indore
Ujjain Road, Indore, M.P.-453111
E-mail: texcon@svvv.edu.in | Website: http//mantrasurat.org
Contact: +91-0261-2323211, 2337062

Supported by

MAN-MADE TEXTILES RESEARCH ASSOCIATION
Near Textile Market, Telephone Exchange, Ring Road, Surat-395002
E-mail: director@mantrasurat.org | Website: http://mantrasurat.org
Contact: +91-0261-2323211, 2337062

ALLIED PUBLISHERS PVT. LTD.

New Delhi • Mumbai • Kolkata • Chennai • Bangalore • Hyderabad

ALLIED PUBLISHERS PRIVATE LIMITED

D-5, Sector-2, **Noida**–201 301
Ph. Nos.: 0120-4320295/2542557/4352866 • E-mail: delhi.books@alliedpublishers.com

17 Chittaranjan Avenue, **Kolkata**–700072
Ph.: 033-22129618 • E-mail: cal.books@alliedpublishers.com

15 J.N. Heredia Marg, Ballard Estate, **Mumbai**–400001
Ph.: 022-42126969 • E-mail: mumbai.books@alliedpublishers.com

No. 25/10, Commander-in-Chief Road, Ethiraj Lane (Next to Post Office)
Egmore, **Chennai**–600008
Ph.: 044-28223938 • E-mail: chennai.books@alliedpublishers.com

P.B. No. 9932, No. 15, 3rd Floor (Next to Vijaya Bank), 5th Cross, Gandhinagar, Karnataka, **Bangalore**–560009, Ph.: 080-41530285 / 22386239
• E-mail: bngl.journals@alliedpublishers.com / apsabng@airtelmail.in

Sri Jayalakshmi Nilayam, No. 3-4-510, 3rd Floor (Above More Super Market)
Barkatpura, **Hyderabad**–500027
Ph.: 040-27551811, 040-27551812 • E-mail: hyd.books@alliedpublishers.com

Website: www.alliedpublishers.com

ISBN: 978-93-90951-39-0

Foreword

Growing in a Turbulent Environment

The new field that has emerged as a result of the combination of textile technology and medical science is known as medical textiles. There are touches of technological and smart textiles everywhere from the hygienic materials and hospital bed sheets, curtains to surgical masks and gowns used in operation theaters, even surgical threads, bandages and artificial bones, ligaments, artificial kidneys and lives. Medical textiles have thus, been playing an important role in medicine in one or the other form.

Some vital features of medical textiles are that textile materials must be non-toxic, resistant to allergens and cancer. Medical textiles must have the ability to disinfect without any structural or chemical charges. Textile materials must be biologically compatible, and capable of anti-microbial activity i.e., must be resistant to micro-organisms. It must have air permeability and waterproofing properties, and should be resistant to acid and alkali. The quality of the fabric should be good, and it should be rewashable as well as reusable.

The types of fibers used in medical textiles are *Commodity fiber* like natural and regenerated fibers (cotton, silk, wood, pulp, viscose), and synthetic fibers like polyester polypropylene, polyethylene, polyamide, PTFA, carbon, glass and silica; *Specialty fiber* like collagen, calcium alginate, chitin and chitosan. Different types of fabrics are used in the medical and healthcare sector, such as woven fabric, non-woven fabric, knitted fabric, and braided. However, most of the medical textiles are made with woven and non-woven fabric.

Indian textile sector is one of the worst hit sectors due to COVID 19 induced crisis. In both demand and supply it has witnessed several hurdles. Due to health issues, necessarily imposed social distancing and stay at home set up induced reduction in house and global textile (export) demand. Similarly, suspended manufacturing activity impacted exports, domestic commodity stocks and the employees adversely. 13–32% reduction in trade volumes in 2020 (WTO, 2020), global growth falling to –3% (IMF, 2020) and different maritime seaborne scenarios ranging from a return to sector average (around 3% p.a.) after 2022 to growth rates falling by 17% by 2024 (Stopford, 2020) was forecasted for textile industry in the United Nations Conference on Trade and Development.

The world at large is still struggling to get its economical footing under continuously evolving COVID 19. Most sectors are still facing losses and lack of sales due to the economic slow-down. However, technical textiles more precisely medical textiles in India have seen a surge in this span of time.

Medical textiles stood as the primary line of protection for health care and other public/emergency/frontline service providers—our first line of defense against COVID 19. Several manufacturers like Welspun, Sutlej have initiated the production of medical textiles. India blessed with abundant raw materials such as cotton, jute and silk along with a strong value chain and affordable labor, satisfied domestic demand. From manufacturing zero personal protective equipment (PPE) kits in March 2020 to manufacturing 2.5 lakh a day in 60 days allowed India to become the second largest manufacturer after China. Today, India produces around 4.5 lakh PPEs and more than 1.5 crore masks every day. It makes Indian medical textile industry the second largest exporter of masks and PPE kits globally.

Consumption of technical textiles in India is only 5–10% in contrast of 30–70% in some of the advanced countries. Henceforth, Industry, Academia and Government of India have taken primary steps to narrow down the gap. National Technical Textiles Mission has initiated a four-year implementation plan with a total outlay of 1,480 crores INR, to place India in a steering position in global technical textile field. Government of India has taken various other initiatives such as allowing 100% FDI. Already, global giants such as Du Pont, Procter & Gamble, Johnson & Johnson, and Kimberly-Clark, among others, have taken benefit of it. To boost domestic market and job scenario, government is also proposing to mandate the use of technical textiles mandatory in ministries and public agencies.

The government has also initiated integrated textile parks, eight centres of excellence, and the Amended Technology Upgradation Fund Scheme. In an indirect mode of encouragement to technical textile application, the Government of India announced in December 2020 the $1.4 trillion national infrastructure budget. The identified projects include energy, road, railway, urban development, irrigation, and health sectors. These sectors can utilize technical textiles extensively. According to a recent Invest India report, the global demand for technical textiles is expected to grow to $220 billion by 2025 from $165 billion in 2018, at a CAGR of 4%. India's technical textiles market shows a growth of 20% from $16.6 billion in 2017-18 to $28.7 billion by 2020-21, according to the Baseline Survey of the technical textile industry by the Ministry of Textiles.

Along with the emerging technical textile field, the Boston Consulting Group has reported that the apparel sector is also showing 27% drop in apparel market from 4,129 billion INR in 2019 to 3,022 billion INR in 2020. Though dented, still apparel market is alive. It has shifted its offline transaction to e-commerce (7 to 10%). Dependence on social media has increased. Some long-term choice shift such as preference to more casual wear with a greater focus on athleisure with a compound annual growth rate (CAGR) of 20-25%

in the next five years. BCG has predicted a healthy growth of 5,781 billion INR by 2024.

To transform the Covid-19 crisis to an opportunity with more vigor, industry and academia need to encourage new ideas, research, and brainstorming collaborations. Having these objectives in view, I am happy that Shri Vaishnav Institute of Textile Technology is organizing TEXCON-2022 on Medical and Apparel Textiles. The proceedings of the conference will be a useful guide to deal with the challenges and provide opportunity for research in future. The conference has an overwhelming response and selected papers will be included in a book, I welcome all of you to actively participate in TEXCON 2022 and learn from each other.

Date: 10.02.2022

Dr. Upinder Dhar
Vice chancellor
Shri Vaishnav Vidyapeeth Vishwavidyalaya,
Indore

Prologue

Challenges and Opportunities in Medical and Apparel Textiles

Textile industry is currently pegged at an estimate of USD 150 billion (INR 10,834 billion) with 75% of the total consumption being witnessed in the domestic market, while exports accounted for the remaining 25% of the total market size. India ranked 2nd in textile export with 6% of global share and stood 5th in apparel export with 4% of global share. In terms of employment, Indian textile industry is the second largest employer after agriculture, providing direct and indirect employment to nearly 100 million people in India.

The widespread impact of the COVID-19 which has left no sectors unturned and is expected to decelerate the growth projection of the textile and apparel industry in India, which was once projected to grow and reach USD 220 billion (INR 16,637 billion) by 2025–26 (as per the data released by the Ministry of Textiles). According to the data released by the Confederation of Indian Textile Industry (CITI), the recovery for the domestic market is expected to be quite steep post pandemic. India is not the only country feeling the heat. Despite the setbacks, implementing the following solutions along with continuous evaluation can lead India to be a global textile hub.

1. The INR 3,000 billion collateral free automatic loan for businesses.
2. The Reserve Bank of India has announced a moratorium of up to 3 months on payment of instalments of term loans.
3. The government has announced to pay the entire employee provident fund for establishments with up to 100 employees where 90% of the employees are earning less than Rs. 15,000. This is a huge aid for the labour intensive small and medium scale apparel manufacturers.
4. The MSME will provide the benefit to ~4.5 million small businesses, pertaining to 6 months moratorium for interests and for 4 years repayment period including one-year postponement for MSME segments for COVID-19 loans, which would provide the much-needed upliftment to the textile industry.
5. 100% FDI in the sector under the automatic route is expected to attract USD 140 billion (INR 10,485 billion) foreign investments in the coming years.

6. Scheme's like Integrated Textile Parks (SITP) and Technology Upgradation Fund Scheme (TUFS) to encourage the flow of more private equity and to train the workforce.
7. 25% reduction in TCS (Tax Collected at Source) and TDS (Tax Deducted at Source) to improve liquidity and a marginal reduction of 2% in employer's and employees' contribution to the Employment Provident Fund.
8. Indian exporters face a duty disadvantage of 9.6 per cent in the EU market. To assist Indian exporters, a RoSCTL (rebate of state and central taxes and levies) scheme to assist exporters has been extended. The different schemes announced under the Rs 20 lakh crore post COVID-19 package by the Government will immensely assist exporters. The Rebate of State and Central Taxes & Levies (RoSCTL) on apparel and made ups has been extended for the upcoming financial year until merged with Remission of Duties and Taxes on Exported Products (RoDTEP). This will make the textile sector more competitive and boost

OPPORTUNITY IN MEDICAL TEXTILE SECTOR

ASTM International is providing no-cost public access to important ASTM standards used in the production and testing of personal protective equipment—including face masks, medical gowns, gloves, and hand sanitizers—to support manufacturers, test labs, health care professionals and the general public as they respond to the global COVID-19 public health emergency.

In recent times, hygiene has become the most important factor. PPE kits, masks, disposable towels, disposable bed sheets etc. are new areas where Indian manufacturers can take the lead. For this to happen, ten new laboratories have also been set up for testing of PPE coveralls all over the country.

India has also become the second largest manufacturer of PPE in the world with more than 600 companies in India certified to produce PPE's today whose global market worth is predicted to be over USD 92.5 billion (INR 6,927 billion) by the end of 2025, as compared to USD 52.7 billion (INR 3,971 billion) in 2019 which would further enhance the textile products demand in coming years.

Companies have prepared such a module that world-class PPE kits will be made on a large scale in the country itself. One lakh PPE kits are being prepared every day. According to the estimation, PPE kit will have a market of 10 thousand crore rupees in India in the next one year. The Apparel Export Promotion Council (AEPC) said the sector is gearing up to start producing the medical textiles, with a view to making India a hub for sourcing of PPE kits over the next few years. According to a study, there will be a domestic demand of Rs. 10,000 crore for the next one year and internationally there will be a USD 60 billion business in 2025, whereas India has done only USD 260 million so far last year, AEPC Chairman A Sakthivel said. According to the same study, 58% of textile manufacturers are willing to re-purpose production towards COVID-19 response goods (facemasks, towels, bed linen, and patient gowns).

Demand for Antibacterial and Health monitoring Garments

Research shows that viruses can remain active on textile surfaces for up to two days. Every textile, whether it is clothing wear or home textiles, the most stylish trend diva wear, weaved or woven, grown-ups or children, formal or easy going, will have customers' preference if they have in-built cleanliness inside it.

Attire and home material Industry must concentrate on fabrics that are:

- Anti-microbial/Hostile to viral/Enemies of Microorganisms
- Anti-smell, Hostile to Perspiring
- Anti-pilling (to abstain from capturing of Microbial)
- PU covered Denims

The high demand for health-monitoring wearables will offer immense growth opportunities. To make the most of the opportunities, market vendors should focus more on the growth prospects in the fast-growing segments, while maintaining their positions in the slow-growing segments.

High demand for health-monitoring wearables has been instrumental in driving the growth of the market.

Quality and Sustainability

Many experts believe in increasing quality for increasing profit margins. Also, the reason that developed nations import from China is the quality.COVID-19 has made us aware of our vulnerability and extended lockdowns have been a period of reflection. Consumers have discovered their consciousness and started sorting, re-use and recycling of textiles through innovation, encouraging industrial applications and regulatory measures such as extended producer responsibility will be of immense help.

The sustainability strategy will aim at strengthening industrial competitiveness and innovation in the sector, boosting the Indian market for sustainable and circular textiles, including the market for textile reuse, addressing fast fashion, and driving new business models. One of the long-lasting changing consumer behaviors is social distancing and the preference for sanitized products. Moreover, the most critical part of an apparel purchase in a market like India is trial. The world will not be the same after COVID-19. For the textile industry it is very important to recover the trust of consumers. Only through sanitizing, brands will be able to speed up regaining consumers' trust, guaranteeing the fast recovery of our industry.

This crisis has shown us how interdependent the world is. In India, The textiles sector is gearing up to make the 'aatmanirbhar (self-reliance)' path a success. We promote self- supporting networks and creators empower 'Handcrafted in India', 'Make in India', 'Purchase Indian and advance Indian'. As far as demand and supply is concerned, out of $140 billion textile only $30 to $35 billion is for exports the remaining $100 to $105 billion is for domestic industry. We have a huge potential and such a big domestic market hence we are not dependent on exports alone.

Shri Vaishnav institute of textile technology (SVITT) a constituent institution of SVVV. Indore has thus initiated a national level conference TEXCON–2022 with the theme "Challenges and Opportunities of Medical and Apparel Textiles" focusing the latest advancements in the Medical textiles field. The conference is aimed to provide a platform to bring academia, research and industry together to contribute their intellectual ideas and their works. We would like to thank all authors for their valued contributions. We would also thank the conveners, editorial team and members of organizing team of TEXCON–2022. We acknowledge the support received from faculty and staff members of SVITT AND SVVV Indore.

Date: 21.02.2022

Dr. V.R. Sampath
Director
Shri Vaishnav Institute of Textile Technology
SVVV, Indore

About the Editors

Dr. Namit Gupta is Professor of Electronics and Communication Engineering and Director of Shri Vaishnav Institute of Technology & Science, Shri Vaishnav Vidyapeeth Vishwavidyalaya, Indore. His areas of research are Quantum Computing, Embedded System, Spintronics, 5G Communication System etc. Dr. Gupta has done his B.Tech in Electronics and Communication and M.Tech in Embedded System from DAVV Indore. He has done his Ph.D in Quantum Cellular Automata based design from DAVV Indore. He has published and presented more than 50 research papers in various National and International Journals and Conferences. He has been awarded one patent in his name. Two candidate have been awarded Ph.D under his supervision and presently 7 other scholars are pursuing Ph.D under his supervision.

Dr. V.R. Sampath is Professor of Textile Engineering and Director of Shri Vaishnav Institute of Textile Technology, Shri Vaishnav Vidyapeeth Vishwavidyalaya, Indore. He has 22 years of academic and 17 years of industrial experience. His areas of research are Compact spinning, Process control in spinning and weaving, Technical textiles, Textile composites etc. Dr. Sampat has done his B.Tech in Textile Engineering from the University of Madras. He has done his M.Tech and Ph.D in Textile Technology from Bharathiar University, Coimbatore. He has published and presented more than 60 research papers in various National and International Journals and Conferences. He has been awarded one patent in his name. Seven candidate have been awarded Ph.D under his supervision. He has to his credit of more than 1 Crore INR worth UNDP research projects with Govt. of India and has successfully executed and completed the projects.

Prof. Tapan Kumar Sinha is Professor of Shri Vaishnav Institute of Textile Technology under the umbrella of Shri Vaishnav Vidyapeeth Vishwa vidyalaya, Indore. He has wide 16 years of industrial experience as Manager (R&D) and there after 21 years of Teaching Experience in the field of Textile Technology. He obtained M. Tech in Textile Engineering from IIT Delhi and presently pursuing PhD from SVVV, Indore. His specialization is in textile engineering in the field of yarn manufacturing and machineries. Besides taking various industrial consulting assignments, he has the wide experience in technical assignments to the various textile Industry in the country as well as in the abroad for problem solving, technical seminar and Exhibitions. He has authored/co-authored more than 31 papers and a number of presentations all over the country. He has supervised more than 80 undergraduate and 02

M. Tech dissertations. He is the Chairman of Board of Studies of Textile Engineering of Rajiv Gandhi Proudyogiki Vishwavidyalaya, Bhopal and also in the Shri Vaishnav Vidyapeeth Vishwavidyalaya, Indore. Prof. T.K. Sinha was awarded by Institute of Engineers in 2014 for his lifetime achievement in developing and spreading the Textile engineering education in the state of Madhya Pradesh and by Textile association of India, M.P. unit in the year of 2015. He is also a member as subject expert in the textile goods purchasing committee of DAVV University. His areas of interest are yarn manufacturing, textile testing, fibre science and technical textile. He is also a life member of Textile association oflndia.

Dr. Shamayita Patra is working as Associate Professor in Shri Vaishnav Institute of Textile Technology at Shri Vaishnav Vidyapeeth Vishwavidyalaya, Indore, India. She has 5 years of Teaching and 8 years of research experience. She has completed her B.Tech in Jute and Fibre Technology from University of Calcutta. She earned her M.Tech in Fibre Science and Technology and Ph.D in Biomedical Engineering from Indian Institute of Technology Delhi (IIT Delhi). She was Dr. D.S. Kothari Postdoctoral fellow from 2013 to 2016. Her areas of research are Medical Textile, Nanomaterials, Soft Tissue Engineering, Natural Fibre, Textile Engineering etc. She has published and presented more than 25 research papers in various National and International Journals and Conferences. She has been awarded two patents and published one patent. Under her supervision presently 5 scholars are pursuing Ph.D research. She has been awarded with IEI Young Engineers Award 2018–19 in the field of Textile Engineering by Institute of Engineers (India) and Young Scientist Award by IARDO in 2019.

Dr. Syamal Maiti completed his M.Tech and Ph.D from IIT Delhi. Currently he is working as Assistant Director at Man Made Textiles Research Association (MANTRA), Surat. He has 2 years of teaching, 2 years of Industrial and 7 years of research experience. Currently he is working on nonwoven and technical textiles, smart textile, etc. He has published and presented more than 10 research papers in various National and International Journals and Conferences. He has been awarded one patent.

Contents

Extraction and Analysis of Fibres from Cornhusk: A Sustainable Approach

Aadhar Mandot, Savan P. Chokshi and Sanjay B. Bambhaniya*
Department of Textile Engineering,
The Maharaja Sayajirao University of Baroda, Gujarat, India
E-mail: *sanjubambhaniya06@gmail.com

ABSTRACT: *Natural fibre crops are gaining more importance to human beings concerning their usefulness. Most of the natural lingo cellulosic fibres with their unique properties are called the fibres for the future. In India, large quantities of corns are cultivated, in which corns are being used and its remaining portions known as corn husk is discarded as waste. These corn husks are eco-friendly and bio-degradable and are mainly used for cattle feeding, artificial sugar, paper, and fuel. The aim of this work is to extract fibres from corn husk which would contribute to the need for natural and bio-degradable material of today's textile society.*

In this work, different chemicals with varying concentrations were used to extract fibres from the corn husk. The extracted fibres were investigated for properties like density, diameter, fineness and tensile. The structural analysis of extracted fibre was also carried out using FTIR, XRD, and capillary action.

Keywords: Cornhusk, Extraction, Biodegradability, Tensile Properties, Lignocellulose.

1. INTRODUCTION

Maize is grown under diversified climatic and soil conditions. Maize is mainly a rainfed Kharif crop. Large quantities of corn are being cultivated all over the world which is mainly used to produce starch, paper, fuel and xylitol (artificial sugar) apart from food. India ranks 7th in corn production which comes to 27.8 MT according to IIMR for the year 2018–19. [1] Corn husks are the part that covers and protect the ears of corn. It is the strongest part of a corn plant and its nature is to protect the seeds. Cornhusk is the waste part of maize. It is eco-friendly and bio-degradable. Annually in India, thousands of tons of waste is being collected, whereas about 1000 MT of waste is collected in the year 2020–21. [2] The use of husk provides a highly economical source of raw material for producing textile products. [3] Corn husk has a limited source of research. It is composed of cellulose (42%), lignin (13%), ash (4.2%) and other materials (41%). [4,5] Cellulose fibres are interconnected through lignin and hemicellulose to form large bundles micrometres to millimetre range in corn husk. Ligno cellulose fibres are utilized in various areas such as automotive, building and packaging applications to replace the conventional man-made fibres. Additionally, utilizing this would provide environmental benefits as the number of by-products being disposed of would be reduced. [3]

In this study, fibres are extracted from corn husk by different alkali with different varied concentrations and time. The effects of extraction parameters on the physical properties of fibres like density, diameter, fineness, and tensile strength have been investigated. Structural analysis of extracted fibre was also carried out using FTIR, XRD, and capillary action.

2. MATERIALS AND METHODS

2.1 Materials

Corn husks were procured from the commercial market. Glacial acetic acid, Caustic soda, Methanol, Turkey Red Oil (TRO) and Dyes used for this research was supported by the department of textile processing at The Maharaja Sayajirao University of Baroda.

2.2 Methods

The fibre extraction process was carried out by subjecting 100 grams of cornhusk in a chemical bath containing NaOH at different concentrations. The bath was heated for 20 minutes at 80°C, further, it was washed with fresh water. At the end of the process, 2% TRO was added to the fibres and dried at room temperature. After drying, combing of the cornhusk was carried out to extract the fibres from it.

Table 1: Treatment of Cornhusk with Different Acids and Base

Sr. No.	*Chemical Used*	*Duration*
a.	15% NaOH	24 hrs
b.	15% glacial acetic acid	24 hrs
c.	Oven dry 450°C	45 seconds
d.	20 ml methanol	24 hrs

Sr. No.	*Chemical Used*	*Duration*
e.	10% NaOH	24 hrs
f.	11.11% glacial acetic acid	24 hrs
g.	50 ml methanol	24 hrs

3. RESULTS AND DISCUSSIONS

Table 1 shows the treatment of cornhusk with different acids and bases. As shown in Figure 1, it can evident that the cornhusk was less affected at low acid/base (except NaOH) concentration. At low concentration of the gummy material was not fully removed. Whereas at higher concentrations the Corn husk fibre along with gummy material gets dissolved. At the same time, when the material was treated with a higher concentration of NaOH, the fibres were broken into parts and not dissolved into NaOH like other chemicals as shown in Figure 2a. Further, the cornhusks were treated with moderate concentrations of NaOH like 0.25, 1.25, etc., which gave optimized results as shown in Figure 2 b&c. Finally, cornhusks were further treated with different concentrations of NaOH at 60°C for 45 minutes as shown in Table 2.

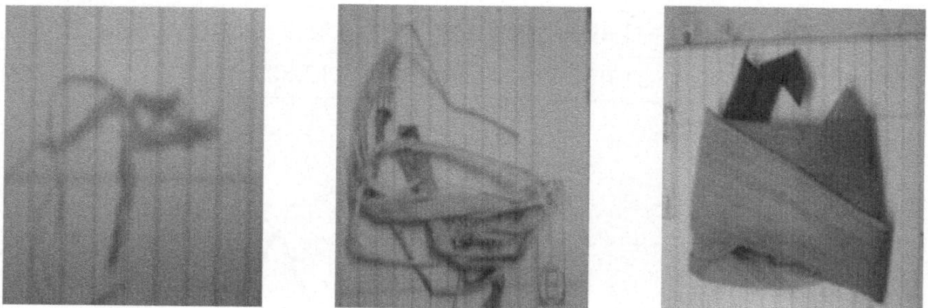

(a) (b) (c)

Fig. 1: (a) 15% Glacial Acetic Acid, (b) 20 ml Methanol, (c) 11% Glacial Acetic Acid

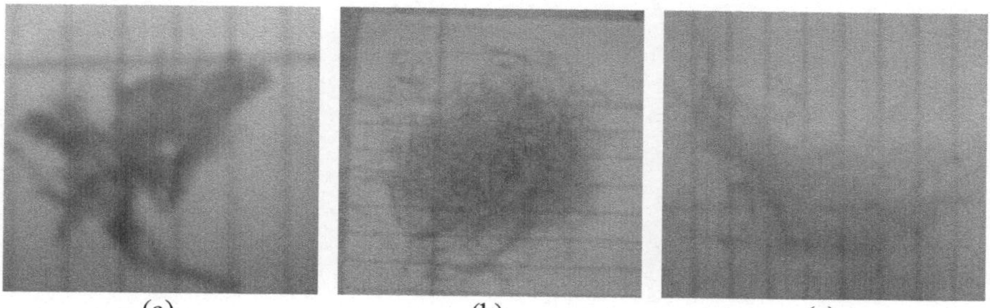

(a) (b) (c)

Fig. 2: (a) 15% NaOH, (b) 0.25%NaOH, (c) 1.25% NaOH

3.1 Physical Properties of Fibres

The extracted fibres were assessed for their properties such as linear density (Tex), diameter and tensile strength. Readings of each prepared sample were carried out at standard testing conditions and its mean, %CV was calculated. The density of fibre was obtained using a gradient tube, which is 1.79 g/cc. The values obtained from the respective test are mentioned in Table 2. Properties were the obtained yield percentage.

Table 2: Tex, Diameter & Yield % Age by Treating Corn Husk with NaOH of Different Concentration

Sample. No.	*NaOH conc. (%)*	*Tex*	*Mean Diameter (mm)*	*CV Diameter (%)*	*CV Tex (%)*	*Yield (%)*
H1	00.25	43.66	00.33	45.10	30.67	–
H2	00.75	24.78	00.31	23.42	42.28	00.40
H3	01.00	30.63	00.25	10.40	27.49	04.64
H4	01.25	28.13	00.31	30.57	22.90	04.66
H5	01.50	24.36	00.26	01.90	24.06	01.81
H6	01.75	23.89	00.29	45.90	09.68	06.23
H7	02.00	31.03	00.27	12.80	35.97	02.60

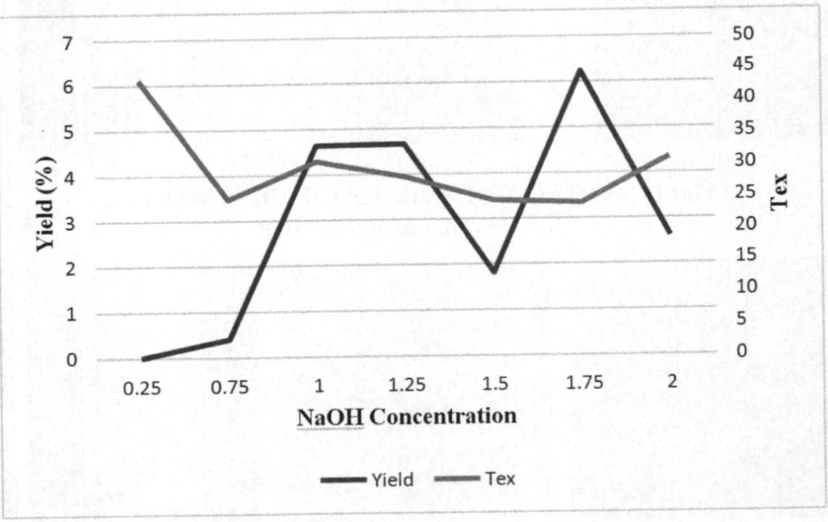

Fig. 3: Relation Between %yield and Tex of Cornhusk Fibres at Different Concentrations

Figure 3 shows the relation between %yield and linear density (tex) of cornhusk fibres at different concentrations. It can be seen that at lower concentrations almost a similar effect was observed to other acids and bases. As at lower concentrations fibres would be affected less so the material would not be extracted in fibre form and hence the tex value would be higher. At 0.75% NaOH concentration fibres were extracted with lower %yield and with lower tex value compared to 0.25% NaOH concentration. On analysing the tex value, %yield and appearance, optimum concentration was defined to extract fibres as 1.25% NaOH. At this concentration, a better %yield and slightly finer fibres were obtained with a pleasing appearance. Also, the diameter of fibres was tested and was found in the range of 0.25–0.35 mm. Variation in results of diameter and tex value of extracted fibres at different concentrations was significant.

Table 3: Tensile Properties of Extracted Cornhusk Fibres

Sample No.	*NaOH conc. (%)*	*Mean Strength (Gms)*	*Mean Extension (Gms)*	*CV strength (%)*	*CV Extension (%)*
H1	0.25	405.845	1.881	39.249	31.707
H2	0.75	449.59	2.670	42.443	23.211
H3	1	200.97	3.500	30.989	23.211
H4	1.25	316.14	3.720	41.993	24.572
H5	1.5	216.960	3.680	46.162	88.579
H6	1.75	312.98	2.780	19.514	18.952
H7	2	164.23	2.520	33.33	35.837

The extracted fibres were analysed for their tensile properties by Instron Tensile Tester. The results are reported in Table 3. As shown in Figure 4, at lower NaOH concentration higher strength was obtained as the material were not much affected, but the extension was lower. Also, at higher concentrations, lower values of strength and extension were observed as most of the material gets dissolved. At optimum concentration, a good value of tensile strength and extension was observed. However, at some concentrations where the extracted fibres have shown higher strength were having lower extension values and vice versa. At 1.25% NaOH concentration, satisfactory tensile strength and extension values were obtained. Therefore, 1.25% NaOH is considered as the optimum concentration to extract the cornhusk fibres. However, the coefficient of variation for tensile strength and extension was in the higher side.

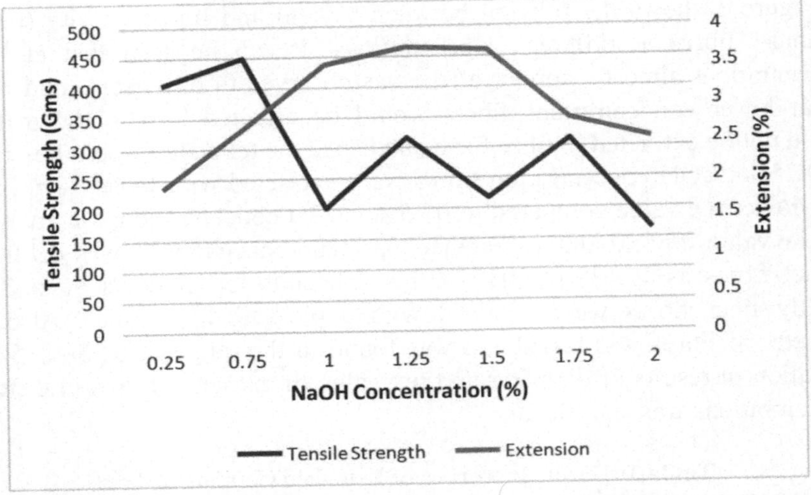

Fig. 4: Tensile Properties of Cornhusk Fibres

3.2 Structural Analysis of Fibres

3.2.1 *Fourier Transform Infrared (FT-IR) Spectroscopy Analysis*

Figure 5 shows the FT-IR spectra of cornhusk fibre which was conducted on Shimadzu spectrometer. A strong peak of –OH stretching was obtained at 3628 cm^{-1}. Stretching of $-CH_2$ was observed at 1478 cm^{-1}. A strong vibration peak of –COO was obtained at 1334.73 cm^{-1}. A vibration peak for C–O–H was found in the spectra range of 450–650 cm^{-1}. Also, the strong peaks of ketones i.e. C–C=O was observed in the spectra range of 560–510 cm^{-1}. The FT-IR spectra confirm the presence of important groups for cornhusk fibre.

3.2.2 *X-Ray Diffraction (XRD) Analysis*

The crystalline nature of cornhusk fibre was studied with the aid of X-ray diffraction as shown in Figure 6. For cellulose, peaks were scattering angle (degrees 2θ) observed at 16° to 22° and 32° to 35° 2θ. The crystallinity and crystallinity index of the corn husk fibre was 99.65% and 7.27% respectively. However, removal the non-cellulosic constituents of the fibres by chemical treatment may change the degree of crystallinity and crystallinity index. Treatment with alkali leads to the removal of adhesive materials like lignin, hemicellulose, pectin, etc. During chemical treatment, with the dissolution of the cementing materials (matrix) will be dissolved, and the remaining pure crystalline particles get isolated. They show increase in orientation along a particular axis, due to their similarity in shape.

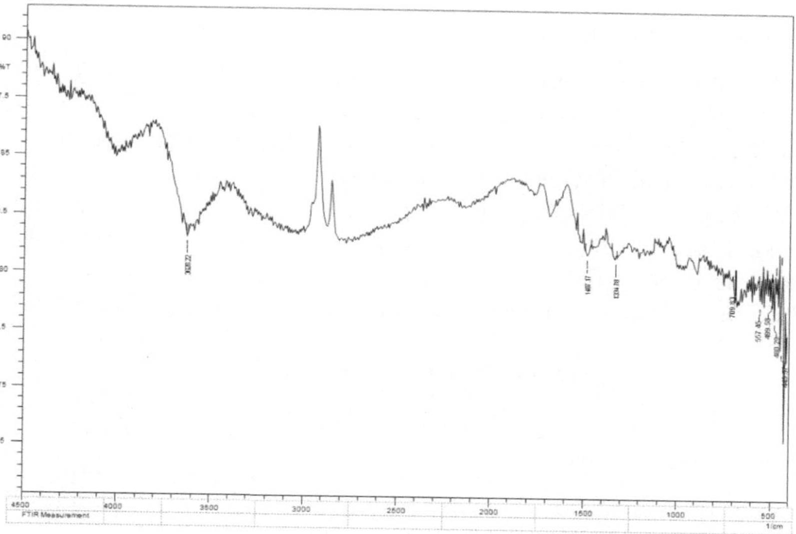

Fig. 5: FT-IR Spectra of Cornhusk Fibre

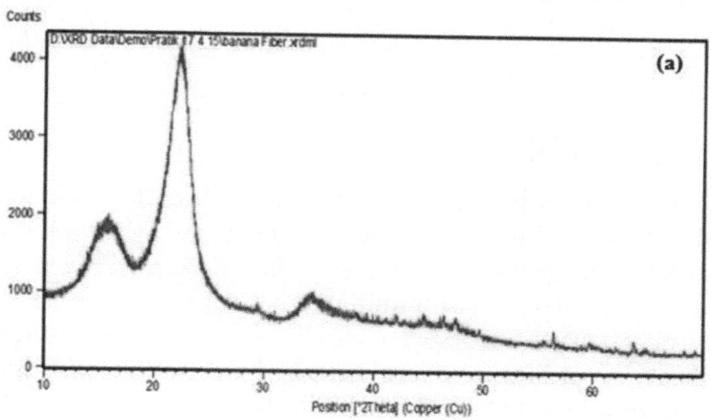

Fig. 6: XRD Analysis of Cornhusk Fibre

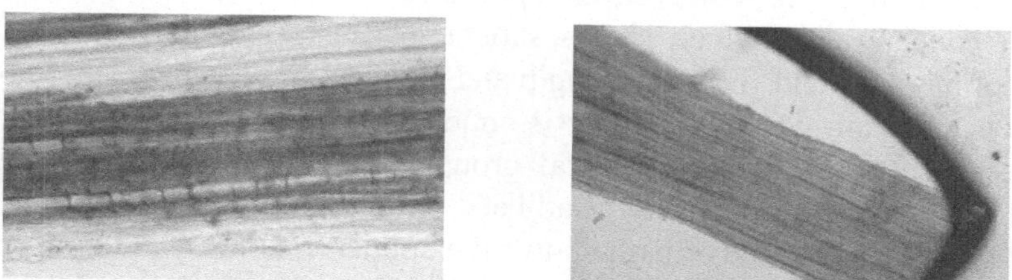

(a) (b)

Fig. 7: (a) Microscopic View of Undyed Sample,
(b) Microscopic View of a Dyed Sample

3.2.3 *Capillary Action*

Capillary action was studied by suspending a single fibre in two dye baths which are prepared by adding 5 ml of respective dye in 50 ml of water. The results of capillary action have been dictated in Table 4, which shows a sudden rise of dye at the start which was only a slight after some duration in both the dye bath. Figure 7 shows the microscopic view of a dyed and undyed sample of cornhusk fibre. It can be observed that dye has occupied the void spaces between the internal fibrils which proves the wicking property of fibres. The structure shows that a single fibre has a parallel arrangement of fibrils.

Table 4: Results of Capillary Testing

	Time (min)	*mm rise*
Bath 1 (Congo Red)	5	2
	10	14
	15	19
Bath 2 (Turqoise Blue)	5	8
	10	14
	15	17

4. CONCLUSIONS

Cornhusk fibre being a lignocellulosic material, with their unique properties are considered as the important fibre for the future, which is discarded as waste. The eco-friendly and bio-degradable fibres were extracted successfully from the cornhusk using NaOH at different concentrations. Other alkalis used in the experiments have not shown enough compatibility to extract the fibres from cornhusk. As at higher concentrations, the cornhusk was dissolved and at lower concentrations it was not much affected.

Treatment of cornhusk with NaOH at higher concentrations has adverse effect on the material. However, the optimized results were obtained at 1.25% NaOH concentration which shows satisfactory values of tex and diameter with optimum %yield. Tensile strength and elongation were also better good at this concentration. Structural analysis conducted by FT-IR and XRD confirm the presence of important chemical groups of lignocellulosic fibres and a crystalline region respectively. Capillary action testing conducted by suspending single fibre in dye bath shows the sudden rise of dye take up which was only slight after some duration. Microscopic view shows parallel arrangement of fibrils in single fibre, the void spaces of which was occupied by the dye that confirms the wicking property of fibres.

REFERENCES

[1] Maize, *Wikipedia.* https://en.wikipedia.org/wiki/Maize#Names.

[2] Raypura, A. (2014). "University of Nebraska Engineers Creating Textiles from Corn Husks". https://sustainablebrands.com/read/chemistry-materials-packaging/university-of-nebraska-engineers-creating-textiles-from-corn-husks, 2014.

[3] Fiscal, R.R. and Dandan, K.B.V. (2016). "Development and Evaluation of Paper from Corn Husks (Zea mays L.) and Snake Plant Fibers (Sansevieria zeylanica)". *International Journal of Science and Research,* 5(8), 867–870. doi: 10.21275/v5i8.3081601.

[4] Yilmaz, N.D. (2013). "Effect of chemical extraction parameters on corn husk fibres characteristics". *Indian Journal of Fibre & Textile Research,* 38(1), 29–34.

[5] Shah, H. and Yang, Y. (2008). "Chemically Extracted Cornhusk Fibres as Reinforcement in Light-Weight Poly (porpylene) Composites". *Macromolecular Materials and Engineering*, 3(293), 235–243.

Melt Spinning of HDPE Fiber using Alternative Route

Ramnath Kumar

Man Made Textile Research Association, Ring Road, Surat, India
E-mail: krramnath1995@gmail.com

ABSTRACT: *In this work HDPE, plasticizer, antioxidant was used to improve the spinnability and drawability of HDPE polymer by sustainable alternative approach of melt spinning to overcome the drawbacks of so-called gel spinning of virgin HDPE, i.e. they are not fully dissolved or molten, unfavorably low polymer concentration in highly flammable solvent and low production rates with environment concerns of solvent used. And melt spinning of HDPE only limited to coarser and low strength fiber. The stearic acid interferes in this polymer chips to provide lubrication properties of HDPE/SA blend that result increase in spinnability and drawability of blend filament upto (5 w/w%) stearic acid concentration. At higher concentration of stearic acid it reduces the chain entanglement of HDPE polymer. Moreover, stearic acid uniformly interacts between polymer chains that result more uniform surface HDPE/SA blend filament. Show HDPE/SA blend filament can produce to fine and comparatively high strength fiber upto 4 gpd tenacity which is reported at 5% stearic acid (w/w% of HDPE). Additional advantage of stearic acid is zero toxicity and user and environment friendly solvent and very low cost. And the HDPE/SA blend filaments can be used for making carpet yarn, fishing nets, ropes and nonwoven fabric.*

Keywords: Gel Spinning, Melt Spinning HDPE, Stearic Acid, HDPE/SA Blend, Spin Ability, Carpet Yarn, Fishing Nets.

1. INTRODUCTION

Scientifically the word polyethylene means: "repeating units of ethylene". They are invisible small parts of ethene monomer which is constituent for polyethene over production. Thermoplastic polymer polyethene (PE), could be melted and remolded than can recovered to a solid state of matter. Suggested chemical name of polyethylene was basically derived from the chemical derivations of monomer. Polyethylene was produced by molecules that have long chains of ethene monomer. It could be produced through different polymerization method such as radical polymerization, anionic addition polymerization, ion coordination polymerization or cationic addition polymerization, because ethylene has not any substituent groups that affect the stability of the polymer propagation head and different polymerization methods gives different type and different grades of polyethylene.

Most common polymer used for melt spinning are PP, PE, PET, Nylon 6, Nylon 6, 6. Though, melting is the essential criteria to produce filament through melt spinning technique, but it is not the sufficient condition. However, gel spinning is carried out only those polymers which have very high molecular weight so that to reduce chain end effect (i.e. to reduce defect), chain of the polymer is linear so that to achieve extended chain structure and minimum chain interaction between chain for high draw ability. [1-5]

2. MATERIALS AND METHODS

For the work, High density polyethylene (HDPE) was purchased from GAIL India limited having grade I60A080, Extrusion molded. And SA and antioxidant (Irganox 1010, Irgafos 168) was purchased from the Sigma Aldrich co. Ltd.

The specifications of all the materials are described in Table 1.

Table 1: Parameters of Materials Used

Material	*Parameter*	
	Density (g/cc)	*Melting Temperature (°C)*
HDPE	0.952	130
Stearic acid	0.85	69
Irganox 1010	1.07	116
Irgafos 168	0.98	182

First of all Stearic acid and antioxidant (0.05% wt. of stearic acid) melt solvent was prepared by heated in round bottomed flask up to the temperature 90°C with continuous nitrogen purging and mechanical mixing at 300 rpm for 3 hr. Prepared solution of above solvent was taken in the 2%, 3%, 5%, 10%, 15% and 20% (w/w) of HDPE and then it is mixed with HDPE chips by constant stirring at 90°C. And they were made into different sample of HDPE/stearic acid followed by spinning in lab scale DSM Xplore micro Twin screw extruder (Shown in Figure 1.) with following set of parameters:

- Processing range temperature:160–200°C
- RPM: 50 under nitrogen blanket
- Dwell time: 3–5 min to prepare extruded stand of 1.5 mm diameter and it is cooled in air to Room Temperature.

Fig. 1: DSM X-PLORE Twin Screw Extruder

To prepare five different sample name as HDPE, HDPE/SA 5, HDPE/SA10, HDPE/SA15, HDPE/SA 20.

The as spun filament are further drawn on two stage drawing machine to remove natural draw ratio, improve modulus of filament and reduce the extension of filament.

2.1 Characterization of Filaments

1. The prepared filaments were characterized by
2. Fourier transform infrared spectroscopy (FTIR) Analysis,
3. SEM of blend filament surface,
4. Tensile testing,
5. Thermal analysis and
6. Wide Angle XRD Analysis.

3. RESULTS AND DISCUSSIONS

3.1 Throughput and Draw Ratio of Blend Filament

Throughput of blend filament (HDPE/SA) was measured for all the sample shown in Table 2.

Table 2: Throughput and Draw Ratio (DR) Relation at Different Concentration of SA

Stearic Acid (Wt. %)	*Filament Type*	*Throughput (g/min)*	*DR*
0	HDPE	1	3
2	HDPE/SA 2	1.04	5
3	HDPE/SA 3	1.06	7
5	HDPE/SA 5	1.08	9
10	HDPE/SA 10	1.12	8
15	HDPE/SA 15	1.15	6

Table 2 result shows that the throughput of blend was increases with increasing the concentration of SA, this may be due to stearic acid increase the lubrication properties of polymer. And draw ratio of blend filament, first increases with increases in concentration of SA and then decreases after optimum concentration of SA. This is due to stearic acid also decrease the chain entanglement between the polymer chains. So 5% SA concentration can be considered as optimum concentration of SA.

3.2 Tensile Properties

Tensile properties of all samples was carried out to evaluate the mechanical properties of the blend filament (HDPE/SA), for HDPE filament, HDPE/SA blend filament at various concentration of SA has shown in Table 3, the maximum breaking load reported in HDPE/SA (5 wt.% of stearic acid) blend filament due to more draw ratio and subsequently it is decreases with rising composition of SA due to decrease drawability of filament at more concentration of SA. Thus, the mechanical properties of filament are first increases with increasing concentration of SA in HDPE/SA blend filament and then start decreasing after certain concentration of stearic acid it may be called optimum concentration of stearic acid whereby which it shows better mechanical properties of the blend filament at that concentration. The breaking load of HDPE/SA filament have been found to be increasing with increase in draw ratios that was observed in all the sample this is due to increase in orientation of chain with drawing of the filament and this has been highest incase of HDPE/SA (5 wt. %) filaments.

Table 3: Mechanical Properties of Blend Filament

Filament Type	*Tenacity (gpd)*	*Modulus (gpd)*	*Elongation. (%)*
HDPE	1	25	68
HDPE/SA 5	4.26	36	25
HDPE/SA 10	2	28	10
HDPE/SA 15	1.6	29	8.5
HDPE/SA 20	0.8	0.85	26

3.3 DSC Analysis

Differential Scanning calorimetry analysis of undrawn and drawn sample of HDPE/SA blend filament (HDPE/SA 5 DR 0, HDPE/SA 5 DR 9, HDPE/SA 10 DR 0 & HDPE/SA 10 DR 8) was carried out at 10°C/min heating rate for one complete heating cycle shown in Figure.2.

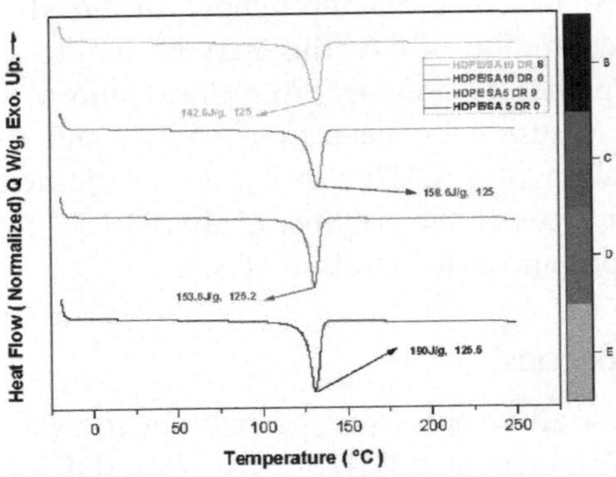

Fig. 2: DSC Heating Cycle of Different Blend Filament

The DSC peak ensures that there is no significant difference between undrawn and drawn sample of HDPE/SA blend filament was observed this is basically due to no more crystallization happen after drawing of filament due to stearic acid which was slow down the rate of crystallization

3.3 SEM Analysis

SEM analysis of all samples was carried out to investigate the surface characteristics of fiber surfaces shown in Figure 3. It has seen that at lower concentration of stearic acid it is strongly connected to HDPE matrix, therefore it is difficult to come out on the surface of fiber and at higher concentration of stearic acid seems to forms separate phase and comes out on the surface of filament. So, 5% stearic is preferred as the most suitable concentration for HDPE fiber spinning among the different HDPE/SA (0, 10, 15, 20 wt. %) blend filament.

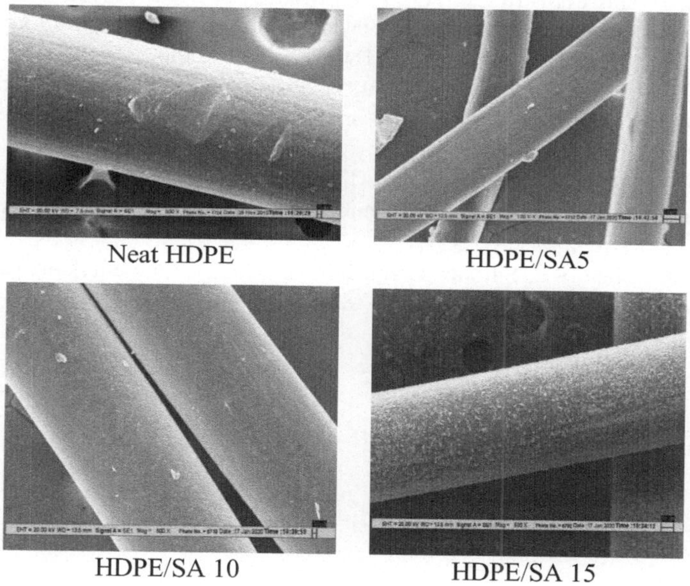

Fig. 3: SEM Images

4. CONCLUSION

Spinning of HDPE fiber has been possible by using alternative route of melt spinning using plasticizer/solvent such as stearic acid, peanut oil etc.. During extrusion process of HDPE/SA blends, the small drops of stearic acid trapped in HDPE has been observed. And during process of quenching of fiber, interaction between HDPE and stearic acidobserved also. The draw ratio of neat HDPE has been achieved upto 3 only because have limited extent draw ability and low extension at break of HDPE, but in case of HDPE/SA (5 wt. %) filament Draw Ratio have been achieved to 9. In case of HDPE/SA blend. Tenacity (gf/tex) of HDPE/SA (5wt. %) filament have been found to 4.26 gpd, with respect to neat HDPE filament about 1gpd. Also, at lower concentration of stearic acid around 2–3 % (w/w% of HDPE) draw ratio reported to 5 only. So maximum draw ability and better mechanical properties of HDPE/SA blend filament has been found at 5% SA.

DSC curve of HDPE/SA blend filament shows that endothermic peak almost similar to neat HDPE filament. Also SEM Images shows that satisfactory surfaces of HDPE/SA 5 blend filaments.

REFERENCES

[1] Strom, E., Thomas, R. and Seth, C. (2011). “The History of Polyethylene”. *ACS Symposium Series*, 1080, 115–145.

[2] Smith, P. and Lemstra P.J. (1979). “Ultrahigh-strength polyethylene filaments bySolution spinning/drawing, Influence of solvent on the drawability,” *Macromolecular Chemistry and Physics*, 180(12), 2983–2986.

[3] Kavesh, S. and Prevorsek, D.C. (1985).”Producing high tenacity, high modulus crystalline article such as fiber or film, USA Patent no.4551296, Allied Corporation.

[4] Liang, J.Z. (2019). “Melt spinning flow behavior of high-density polyethylene blended with low-density polyethylene”, *Plastics, Rubber and Composites*, 48(6), 256-262.

[5] Sona, K., Chen, S.I., Shiono, T. and Doi, Y. (1985). “Polyethylene, catalysts and (co-polymerization”. *Polymer*, 26, 1888–1890.

Study on Comfort Aspects of Phase Change Material Coated Nonwoven Interlining

Md. Vaseem Chavhan
Department of Knitwear Design, NIFT, Hyderabad
E-mail: wasim.chavhan@yahoo.com

ABSTRACT: *The phase change materials (PCM) known for its heat releasing and absorbing property with change in the temperature. The PCM are applied to the textile substrate in the different form in order to get the thermo psychological comfort. The PCM can be applied by the padding mangle, spraying and by coating. In apparel the application of the PCM material varied from socks, inner wear to the interlining material. In present study the PCM pest was prepared and applied to the specific region of the nonwoven interlining. The PCM past is applied in the form a circular shape by maintaining a certain distance in between. The material has been tested for the its heat of latent by DSC and the coated material is tested for the stiffness and the permeability. There is good heat absorbency by the PCM coated material was observed but with the loss of the permeability and the flexibility of the fabric.*

Keywords: Phase Change Material, Thermal Comfort, Nonwoven, Interlining.

1. INTRODUCTION

The clothing is the basic need of the human being, which provides the protection, comfort and the aesthetic look. With the lifting up of the living of standard, the expected requirements from the cloth are increasing. There is a more focus on increasing the comfort of the garment assembly by using the different techniques. There are various types of the comfort are there related to the clothing, broadly the clothing comforts are; thermo physiological, tactile and the psychological. Out of that the tactile comfort is depends on the type of fabric is used and the way in which the fabrics are stitched to for the garment. The psychological comfort is depend on the mentality of the particular user. While the thermo physiological comfort is related with the heat and the moisture balance from the body and the atmosphere through the fabric layers. There are various techniques are there to improve the thermo physiological. The thermo physiological can be improve by engineering the fabric assembly by selection appropriate fabric construction parameters and the type of the fiber. The other way to improve the thermo physiological is by the use the coating materials. The various types of the coating material can be used for the purpose like which are capable of absorbing the heat and some other coated material which are able to transfer the moisture vapor and the liquid. Again, the requirement of the thermal comfort is varying based on the climate and the metabolism activity of the person. For clod climate with low activity, it is required to retain the heat by using the insulated material. While in hot atmosphere with more activity it is required to dissipate the heat release by the body to avoid the hyperthermia. In such condition the more permeable fabric can be used or the slits can be given at various place for the heat dissipation. But for the garment where it is required to cover the garment with the less permeable or impermeable material in tropical condition the use of permeability at the fabric layers is not sufficient. In such conditions, the phase change materials (PCM) can be used in the form of coating to provide the thermal comfort by absorbing the heat released by the body. The PCM materials are works on the principle of heat absorption when the changing of phase from liquid to solid and the heat dissipation when the phase of the material is changing from solid to liquid. There are various studies are available on the characterization and the application of the different types of PCM on to the fabric. [1]

In one of the studies, the PCM was applied in the form of the micro-encapsulation on to the different layers of the fabric. The study was then carried out by changing the environment from hot to cold and it was found that, with change in the climate the cooling or heating effect provided by the material is last for the short time of 15 minutes. Further it was found that the on an average 6.5 W decrease in a heat lost from the body for the single layer and 13.2 W for the double layer by PCM coating compared to uncoated fabric layer. [2] In some other work the transient effect of heat lost by the boy was studied when there is a sudden change in the climate from the warm to cold. Here it was founded that the heating effect provided by the heat release by the PCM is last for the 12.5 minutes which is further depends on the propert of the PCM material and the climate conditions. By experimental study it was found that the PCM decreases 40–55W/m^2 of heat release from the body. [3]

On other hand the effect of the change in climate from cold to the hot on to the heat release by the PCM coated fabric was studied. The cooling vest was designed and the mathematical model has been developed. It was found that the faster cooling is possible by proper selection of the melting temperature of the PCM material and further the cooling effect can be increase by increasing the mass of the PCM material. [4] The addition of the PCM material affect the properties of the fabric related to the weight and the permeability of the fabric. The change in the properties permeability and the handle of fabric is depends upon the amount of PCM material used. [5, 6]

In present study the PCM material has been coated to certain area only that is printed in a circular shape. Instead of all surface coating the printing is done to provide the permeable area at the unprinted surface as the permeability will be lost at the PCM coated region.

2. MATERIALS AND METHODS

2.1 Material

The needle punched nonwoven fabric of thickness 1 mm was used as the interning fabric material. The paste was prepared by adding the PCM material in the form of powder and the binder to the water to get the 10% paste concentration. The CRODATHERM Bio-Based Phase Change Material was selected as the PCM material. The stencil with circular shape of 1cm diameter was prepared for the application of PCM paste on the fabric. The prepared PCM paste was applied on to the fabric (non-woven) using the stencil. After application of the paste, it was dried using hot air oven at 80°C followed by curing at a temperature of 130°C The prepared material is checked for its heat absorption. And results should be drawn accordingly for non-woven fabric for air permeability and stiffness.

2.2 Method

The printed stencil is prepared in such a way that a unit are of 3 cm by 3 cm cell one circle of 2 cm diameter covered. The total area occupied by the printed circular region was 34.9% $\{[(\pi\ 2^2/4)/9] \times 100\}$.

The PCM printed interlined fabric was evaluated for the air permeability, stiffness and the heat absorption. The air permeability is measured using the ASTM D737 standard by passing the air at the standard pressure of water column and by noting the corresponding rotameter reading. The fabric stiffness is measured using the Shirley stiffness tester by measuring the fabric bending length ass per ASTM D1388. Further the bending modulus was calculated using the mathematical expression. The heat absorption capability of the material is calculated using the Differential Scanning Calorimetry (DSC), where the heat is from raised from 15° to 60°C.

3. RESULTS AND DISCUSSIONS

The heat absorption capability of the required material is evaluated from the DSC curve as shown in figure 1. Here in this test it was observed that at 30°C the pcm gets melted, the softening is starts at 28°C. Afterwards with further increase in the temperature the crystallization process stars, from the peak of graph it can be seen that the total of 40 mW of heat is release per gram of the PCM material. The total heat absorbed can be calculated for the given PCM printed interline material as:

Total heat absorbed per gram of PCM = 40 mW

The quantity of PCM per unit cell of 9 cm^2 = 40 mg × 0.1= 4 mg

The total heat absorber for unit cell = 0.16 mW per 9 cm^2

The total heat absorber per m^2 of PCM printed fabric = 177.7 W

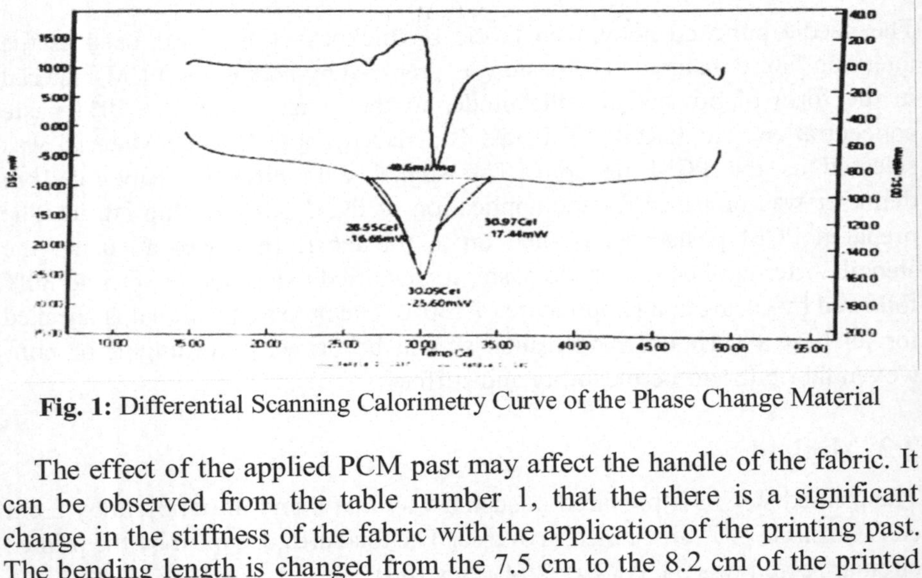

Fig. 1: Differential Scanning Calorimetry Curve of the Phase Change Material

The effect of the applied PCM past may affect the handle of the fabric. It can be observed from the table number 1, that the there is a significant change in the stiffness of the fabric with the application of the printing past. The bending length is changed from the 7.5 cm to the 8.2 cm of the printed PCM material and the correspondingly the bending modus get doubled after printing a PCM paste.

Before treatment:

$$G = M \times C^3 \times 10^3 \text{ mg/cm}$$
$$= 5.080 \times 7.5^3 \times 10^3$$
$$= 2143125 \text{ mg/cm}$$

Bending modulus $= 12 \times G \times 10^{-6}/T^3 = 12 \times 2143125 \times 10^{-6}/1^3$
$= 25.7$ kg/cm

Table 1: The Effect of PCM Printing on to the Stiffness of Fabric

Sl. No.	*Bending Length (cm) Without PCM*	*Bending Length (cm) with PCM*
1.	7.5	8.2
2.	7.5	8.3
3.	7.4	8.3
4.	7.5	8.2
5.	7.4	8.2
Avg. Bending length	7.5	8.2
Bending modulus (kg/cm)	25.7	56.5

After treatment:

$G = M \times C^3 \times 10^3$ mg/cm $= 8.540 \times 8.2^3 \times 10^3$
$= 4708682.72$ mg/cm

Bending modulus $= 12 \times G \times 10^{-6}/T^3$
$= 12 \times 4708682.72 \times 10^{-6}/1^3$
$= 56.5$ kg/cm

It can be concluded that the there is significant change in the stiffness of the fabric the fabric becomes stiffer with the application of the PCM past. Therefore for the applications where the stiffness of the fabric is important it is not suggested to use the PCM in the form of the printed past. But for the applications where the stiffness is not important such as the interlined material, the PCM can be applied there to get the thermal comfort. In present study therefore the suggested application of the nonwoven printed PCM fabric is for the interlined material.

The air permeability of the printed and the non-printed nonwoven fabric was measured as shown in Table 2, to analyse the change and to get the idea relate with the comfort aspect.

Table 2: Air Permeability of the PCM Printed and the Non-Printed Nonwoven Fabric

Sl. No.	*Water Head Column Height (mm)*	*Air Permeability (lts/hr)*							
		Nonwoven Fabric				*PCM Printed Area*			
		Sample 1	*Sample 2*	*Sample 3*	*Average*	*Sample 1*	*Sample 2*	*Sample 3*	*Avg.*
1.	20	800	890	750	813.33	280	160	240	226.66
2.	40	1300	1400	1300	1333.33	380	250	360	330
3.	60	1700	1900	1800	1800	520	350	500	456.66
4.	80	2000	2300	2100	2133.33	700	700	680	693.33
5.	100	2300	2500	2400	2400	790	790	800	793.33

The air permeability was measured at the different water head pressure of 20 mm, 40 mm, 60 mm, 80 mm and 100 mm heights at nonwoven fabric and at the PCM printed area. It can be seen that the at all the pressures there is significant difference in the air permeability at fabric and at the printed area. As, at the present interlined PCM nonwoven fabric contains the only 35% area is covered with the PCM paste, the change in the air permeability is compensated by the remaining open area of 65% of fabric.

4. CONCLUSION

The PCM printed nonwoven interlined material considered for the study shown the good potential of heat absorption for the intrinsic change in the temperature near 30 degree Celsius. The given fabric shows the significant increase in the stiffness which limits its application but as a interline material in the form of the panel there is a scope for its application. The air permeability is changed significantly at the PCM printed region which is compensated by leaving the unprinted portion of the fabric. Over all the given fabric is suitable to use as a interlined material which a good intrinsic heat absorption property and required air permeability.

REFERENCES

[1] Iqbal, K., Khan, A., Sun, D., Ashraf, M., Abdur, A., Safdar, R., Basit, A. and Maqsood, H.S. (2019). "Phase change materials, their synthesis and application in textiles—a review". *The Journal of The Textile Institute*, 110(4), 625–638.

[2] Ghali, K., Ghaddar, N., Harathani, J. and Jones, B. (2004). "Experimental and Numerical Investigation of the Effect of Phase Change Materials on Clothing during Periodic Ventilation". *Textile Research Journal*, 74(3), 205–214, https://doi.org/10.1177/004051750407400304.

[3] Hamdan, H., Ghaddar, N., Ouahrani, D., Ghali, K. and Itani M. (2016). "PCM cooling vest for improving thermal comfort in hot environment". *International Journal of Thermal Sciences*, 102, 154–167, https://doi.org/10.1016/j.ijthermalsci.2015.12.001.

[4] Bryant, Y.G. and Colvin, D.P. (1992). "Fibers with Enhanced, Reversible Thermal Energy Storage Properties". *Techtextil- Symposium*, 2, 1–8.

[5] Pause, B.H. (1993). "Investigations of the Heat Insulation of Protective Textiles with Microencapsulated PCM". *Techtextil- Symposium*, 245, 1–9.

[6] Pause, B.H. (1995). "Development of Heat and Cold Insulating Membrane Structures with phase Change material". *Journal of Coated Fabrics,* 25(7), 59–68.

A Comparative Water Retention Efficiency Study of Ordinary Cotton and Commercially Available Sanitary Napkin Materials

Shamayita Patra* **and Vishakha Yashwant Dharme**

Shri Vaishnav Institute of Textile Technology, SVVV, Indore, India

E-mail: *shamayitapatra@svvv.edu.in

ABSTRACT: *Women due to their physiological makeup undergo menstrual discharge every 28 days. This bodily fluid discharge comprises mainly blood and a little amount tissue. The body undergoing a menstrual cycle requires proper care to prevent any kind of internal or external infection in the vaginal area. Traditionally women administer cotton clothes in menstruation management. However, reuse of the same clothing causes infection, uneasiness in the body.*

First disposable sanitary napkins were made out of cotton fibers and gauge bandages. Research in sanitary napkins has made huge progress in this field since its first commercialization in 1921 in the United States of America. Now sanitary napkin use not only cotton but also super absorbent polymers (SAP) in their composition. This super absorbent enables better absorption of body fluid.

Present study showcases the evolution of sanitary napkin absorbent materials like simple cotton to superabsorbent material in terms of their water retention efficiency (WRE). In this study WRE of virgin cotton (VC), scoured cotton (SC), fibrous absorbing part of sanitary napkin type 1 (FSN1) and fibrous absorbing part infused with SAP granules of sanitary napkin type 2 (FSSN2) were measured. It has been observed that WRE% increased 3.6 times from VC to SC by simple scouring. SC can compete with FSN1 as both of them provides 720±31.62% and 828±30.33% WRE respectively. FSSN2 provides an outstanding WRE% of 1652±76.94. This preliminary study indicates SC as low cost, biodegradable absorbing material for the manufacturing of low cost sanitary napkins.

Keywords: Virgin Cotton (VC), Scoured Cotton (SC), Fibrous Absorbing Part of Sanitary Napkin Type 1 (FSN1), Fibrous Absorbing Part Infused with SAP Granules of Sanitary Napkin Type 2 (FSSN2), Water Retention Efficiency (WRE).

1. INTRODUCTION

Women due to their physiological makeup undergo menstrual discharge for 3–10 days in every 28 days. Menstrual cycle begins at the age of 12 years and last upto 45 years. This bodily fluid discharge comprises mainly blood and a little amount of tissue. The body undergoing through menstrual cycle requires proper care to prevent any kind of internal or external infection in the vaginal area. Traditionally women administer cotton clothes in menstruation management. However, reuse of the same clothing causes infection, uneasiness in the body.

First disposable sanitary napkins were made out of cotton fibers and gauge bandages. Research in sanitary napkins has made huge progress in this field since its first commercialization in 1921 in the United States of America. Now sanitary napkins use not only cotton but also super absorbent polymers (SAP) in their composition. This super absorbent enables better absorption of body fluid. [1]

The first chosen absorbing material for sanitary napkins was cotton.Cotton fiber's ribbon-like convoluted structure makes it feel comfortable. Its moisture regain is 8.5% and it possesses 65–75% degree of crystallinity. These properties made cotton a natural choice for sanitary napkin preparation. The basic structure of cellulose, the major component of cotton, is given below.

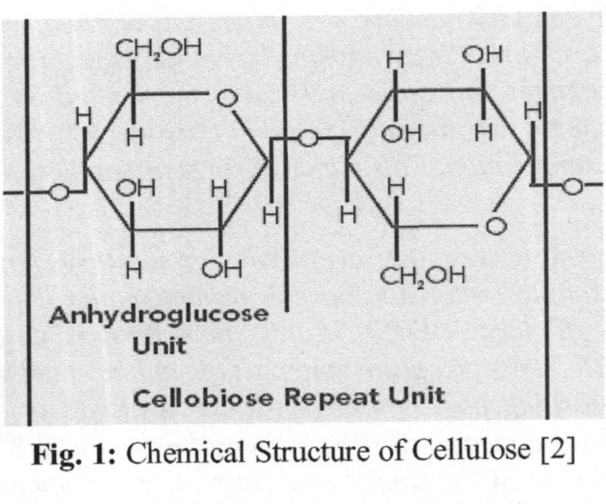

Fig. 1: Chemical Structure of Cellulose [2]

The average blood loss during menstrual cycle is 30–40 ml and maximum of 60 ml. The unmanaged discharge can inflict a bad smell, urinary infection. [3] To combat this situation scientists have developed super absorbent materials. Super absorbent materials are cross linked networks of water soluble polymeric structures. The cross links in the super absorbent materials stop those water soluble polymeric chains from getting dissolved in the water. While water soluble polymers through their hydrophilic functional groups attract and entrap water molecules within the network system. Usually water super absorbent materials are known as hydrogels.

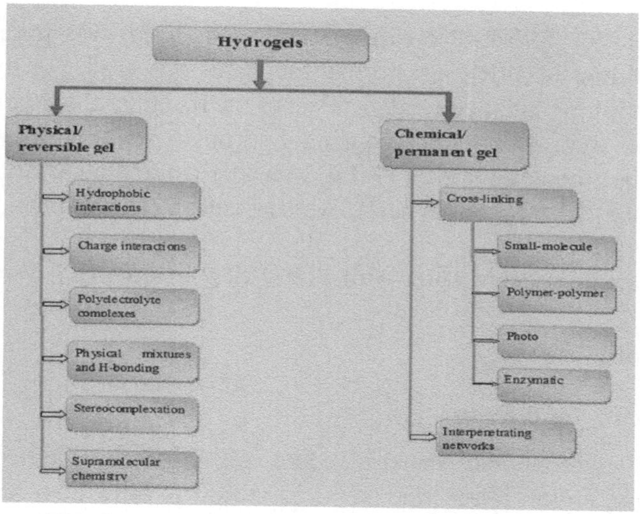

Fig. 2: A Detailed Classification of Hydrogels [2]

The polymeric network system of hydrogels can be synthesized by:

1. *Chemical cross linking*

   Water soluble polymeric structures (like poly vinylalcohol (PVA), gelatin, etc.) can be cross linked by radiation (e.g. γ-ray, UV ray, etc.), and chemical cross linkers (e.g. glutaraldehyde, genipin, enzymes, etc.). Chemical cross linking results in permanent network structure. [4–10]

2. *Physical cross linking*

   Physical cross linking of polymeric chains is dependent on the forces involved in the gel formation such as intra and inter chain hydrophobic, electrostatic and hydrogen bonding. Physical cross linking results in reversible gel structure. Examples of physically cross linked hydrogels are polyethylene glycol-poly(lactide-co-glycolide)-polyethylene glycol (PEG-PLGA-PEG), PLGA-PEG-PLGA, poloxamer (PX), etc. [4, 11, 12]

Hydrogels have found their utilization in health care, agriculture, construction industry and many more other fields. In this study virgin cotton (VC), scoured cotton (SC), fibrous absorbing part of sanitary napkin type 1 (FSN1) and fibrous absorbing part infused with SAP granules of sanitary napkin type 2 (FSSN2) were analyzed to understand their water retention efficiency (WRE) and to identify a cost effective biodegradable substitute absorbing material.

2. MATERIALS AND METHODS

2.1 Materials

In this study VC, FSN1 (Pro-ease), FSSN2 (Stayfree) was purchased from local market. For the WRE test normal tap water was used. For measurement of weight weighing balance CraftsEraVenezzia 10 kg kitchen weighing scale (Thermoplast industries). For scouring hydrogen peroxide (H_2O_2) (make-Bhandari labs), cooking soda (TATA I-shakti) (ingredients: Sodium Bicarbonate refined), detergent (Wheel), wetting agent (Clinic plus, Hindusthan Liver).

For all the experiments water with TDS 180 ppm and pH 7 was used.

2.2 Methods

2.2.1 *Scouring of Cotton*

To remove hydrophobic materials like oils, waxes and other soiling contamination, cotton fibres were scoured at 100°C for 1 hour at pH 11. For scouring 1.6 % (w/v) NaOH, 0.67 (w/v) % detergent, 0.67 (w/v) 5 wetting agent, 1.67 (w/v)% H_2O_2 were used with 1:30 MLR with respect to fibre mass.

2.2.2 *Water Retention Efficiency (WRE) Test*

To calculate the water retention efficiency (WRE) 5 g of dry sample at 27±2°C and 65% r.h. was immersed in the water for 5 minute. Afterwards, the wet sample was hung for 2 minute to remove loosely bound dripping water from the sample. Finally, the sample was weighed. Each sample type was repeated at least for 5 times. The following formula was utilized for calculating WRE:

$$WRE = (Ww - Wd) \times 100/Wd \qquad \dots (1)$$

where, Ww – Wet weight of the sample and Wd – Dry weight of the sample.

3. RESULTS AND DISCUSSIONS

Globally, 1 in every 3 girls face inadequate sanitation and social and cultural problems during their periods. According to DARSA report, every year almost 23 million girls are forced to abandon their studies due to lack of proper menstrual hygiene management. [13] Most affected by menstrual complications lies in between age group of 20–29 year. [14] Awareness and availability of safe sanitation in an affordable price range is extremely important for women and girls. Cost becomes a predominant factor specially for middle, lower middle and lower income class.

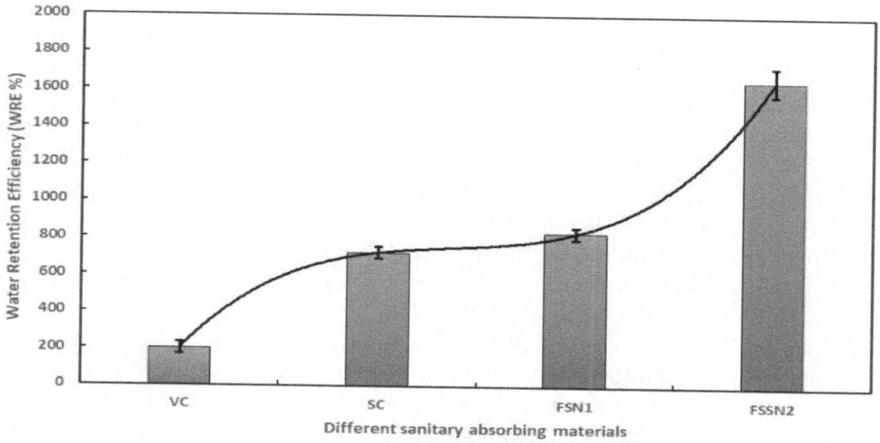

Fig. 3: Water Retention Efficiency (WRE)% of Different Absorbing Materials for Sanitary Napkin Preparation

[Virgin cotton (VC), Scoured cotton (SC), Fibrous absorbing part of sanitary napkin type 1 (FSN1) and Fibrous absorbing part infused with SAP granules of sanitary napkin type 2 (FSSN2)]

In this research WRE of VC, SC, FSN1 and FSSN2 were assessed to identify a cost effective biodegradable alternative absorbing material for the production of sanitary napkin. The WRE values are graphically represented in figure 3. From the Figure 3 it is evident that FSSN 2 has the superior WRE value of 1652 ± 76.94%. Whereas, mere souring of cotton has improved the WRE value of 3.6 fold i.e. from 200 ± 31.62% to 720 ± 31.62%. SC sample provides nearly equivalent WRE that of the commercially available FSN1. FSN1 has a WRE value of 828 ± 30.33%. Hence, SC can be a cost effective and biodegradable alternative to use as the absorbing material for the sanitary napkin.

4. CONCLUSION

WRE% increased 3.6 times from VC to SC by simple scouring. SC can compete with FSN1 due to their almost equal WRE. Though it is not superior as the FSSN2 (mostly biostable), still, SC can be a cost effective biodegradable alternative absorbent material for sanitary napkin production.

REFERENCES

[1] Barman, A., Asagekar, S.D. and Katkar, P. "An Overview on Sanitary Napkins, Technical". Textile.net, https://www.technicaltextile.net/articles/an-overview-on-sanitary-napkins-7850.

[2] "Cotton Morphology and Chemistry". *Cotton for Nonwovens Technical Guide*, https://www.cottoninc.com/quality-products/nonwovens/cotton-fiber-tech-guide/cotton-morphology-and-chemistry.

[3] How Much Blood Do You Lose on Your Period? *Healthline*, 2018. https://www.healthline.com/health/how-much-blood-do-you-lose-on-your-period#:~:text=The%20average%20woman%20loses%2060, than%201%20gallon%20of%20blood.

[4] Pahari, R. (2017). "Cross-Linked Hydrogel for Pharmaceutical Applications: A Review". *Advanced Pharmaceutical Bulletin*, 7(4), 515–530.

[5] Bhattarai, N., Gunn, J. and Zhang, M. (2010). "Chitosan-based hydrogels for controlled, localized drug delivery". *Advanced Drug Delivery Reviews*, 62(1), 83–99. doi: 10.1016/j.addr.2009.07.019

[6] Sung, H.W., Liang, I.L., Chen, C.N., Huang, R.N. and Liang, H.F. (2001). "Stability of a biological tissue fixed with a naturally occurring crosslinking agent (genipin)". *Journal of Biomedical Materials Research,* 55(4), 538–46.

[7] Yoo, H.S. (2007). "Photo-cross-linkable and thermo-responsive hydrogels containing chitosan and pluronic for sustained release of human growth hormone (hGH)". Journal of Biomaterials Science, Polymer Edition, 18(11), 1429–41. doi: 10.1163/156856207782246803.

[8] Liang, H.F., Hong, M.H., Ho, R.M., Chung, C.K., Lin, Y.H., Chen, C.H. and Sung, H.W. (2004). "Novel method using a temperature sensitive polymer (methylcellulose) to thermally gel aqueous alginate as a pH-sensitive hydrogel". *Biomacromolecules,* 5(5), 1917–25. doi: 10.1021/bm049813w

[9] Jin, J., Song, M. and Hourston, D.J. (2004). "Novel chitosan-based films cross-linked by genipin with improved physical properties". *Biomacromolecules*, 5(1), 162–168. doi: 10.1021/bm034286m.

[10] Yung, C.W., Wu, L.Q., Tullman, J.A., Payne, G.F., Bentley, W.E. and Barbari, T.A. (2007). "Transglutaminase crosslinked gelatin as a tissue engineering scaffold". *Journal of Biomedical Materials Research Part A*, 83(4), 1039–46. doi: 10.1002/jbm.a.31431.

[11] Qiao, M., Chen, D., Ma, X. and Liu, Y. (2005). "Injectable biodegradable temperature-responsive PLGA-PEGPLGA copolymers: synthesis and effect of copolymer composition on the drug release from the copolymer based hydrogels". *International Journal of Pharmaceutics,* 294(1–2), 103–12. doi: 10.1016/j.ijpharm.2005.01.017.

[12] Chung, T.W., Lin, S.Y., Liu, D.Z., Tyan, Y.C. and Yang, J.S. (2009). "Sustained release of 5-FU from Poloxamer gels interpenetrated by cross linking chitosan network". *International Journal of Pharmaceutics*, 382(1–2), 39–44. doi: 10.1016/j.ijpharm.2009.07.035.

[13] Tanwar, P., Chakraborty, S. and Apte, K. (2019). "No access to menstrual hygiene is the fifth biggest killer of women in the world". *India Today*, https://www.indiatoday.in/education-today/gk-current-affairs/story/no-access-to-menstrual-hygiene-fifth-biggest-killer-of-women-in-the-world-1552450-2019-06-20.

[14] "Menstrual problems among women in India 2020, by age group". *Statista Research Department*, 2021, https://www.statista.com/statistics/1136538/india-menstrual -problems-among-women-by-age-group.

Future of Textile with Polymer Electrolytes for Wearable Batteries

Amit Saxena and Uttam Sharma*
Department of Physics, Shri Vaishnav Vidyapeeth Vishwavidyalaya, Sawer Road, Indore, M.P., India
E-mail: *druttamsharma1971@gmail.com

ABSTRACT: *With the increasing demand for energy for commercialization and development, there is a need for green energy. Because the traditional sources of energy are draining fast. And to resolve the problem related to these traditional sources of energy, electrochemical devices like solar cells, rechargeable batteries, super capacitors, Li-ion batteries, etc. are the potential candidates. So, there is so much work that has been done in this direction. But still, we have certain limitations with their preparation and installation. In this direction recently the researchers are taken a step ahead to develop flexible and wearable energy storage devices. This would be the breakthrough in the area of energy storage and installation for portable energy storage devices. Here in this work, we have prepared and characterized the polymer-based flexible electrolyte films which offer the conductivity of the order of* 7.95×10^{-5} *S/cm. Also, the other studies related to structural and electrical properties were also analyzed.*

Keywords: Polymer Electrolytes, Conductivity, Flexible Batteries, e-Textile for Energy Storage Devices.

1. INTRODUCTION

In the 21st century, the most intense area of research is green energy. As we all know that the traditional sources of energy are limited for the uses of mankind. The Solar cells, rechargeable batteries, super capacitors, etc. are of prime research interests. But when we are focusing on these areas, then in parallel to development it has numerous limitations as well. Where the foremost is the waste materials of these electrochemical devices. To overcome the limitation nowadays researchers are working on biodegradable materials to produce these electrochemical devices. Latest development in textile is appreciable in terms of e-textile. E-textile is the development of electrochemical devices with the help of textile fibers, which are biodegradable also. In this article, the polymer electrolytes would be discussed in detail with their potential application for textile. The foremost hurdle while developing e-textile materials, which can be wearable, is the rigidity of the electronics device and flexibility of the textile materials. Soft electronics are developed to prepare truly wearable smart textiles in response to this problem. Thus, the device used to power it must be energy efficient in order to minimize the size of the batteries required to run it. It is obvious that an effective, light, flexible and sustainable battery source is necessary to drive all the electronics in smart textile. It would be ideal for the source to be directly embedded in smart textiles during the weaving process as a fiber.

A variety of advances in flexible batteries can broadly be divided into the following four categories:

(a) Flexible conducting polymers (Organics). [1-4]
(b) Flexible fuel cells. [5]
(c) Solar cells (Polymer based) [6-8] and
(d) Mouldable Li polymer batteries. [9-12]

An article published by Bhattacharya *et al.* describes a new type of rechargeable textile battery. [13] Here, the conductive polymeric coating was applied directly over the conductive yarns woven into a textile substrate. Recently researchers have reported the preparation and production of flexible and stretchable electronic circuits, which can be used for various portable electronic devices. [14] In addition to this, the printable highly viscous electrolytes/conducting ink for nonwoven fabrics were also reported in the literature. [15] Here conducting, flexible, and porous textile materials have been developed by "dipping and drying technique with the use of SWCNT (Single-Walled Carbon Nano tubes). [16, 17] They can be useful for the Li-Ion batteries, supercapacitor, etc. Out of all the available types of batteries, Li-ion batteries have attracted the attention of marketers and researchers due to their potential application for electrical vehicles. To improve battery design, reliability, safety, and flexibility, solid polymer electrolytes are used in battery. They can act as an electrolyte as well as a separator. [18]

2. EXPERIMENTAL

"Solution Cast" was used to prepare technique flexible composite polymer electrolyte films. Films were of the order of 200 μm to 300 μmin thickness. [19] Laboratory made fulleren was used to disperse the carbon base material into the polymer matrix. Here to disperse the carbon base material into the polymer matrix, the fullerene was prepared in the laboratory. For the synthesis of this fullerene, the soot is collected by burning phenolphthalein, which is dissolved in benzene, which is continuously stirred over a period of approximately 5 hours and filtered finally. As an end product we obtained Wine-coloured solution. The final confirmation of C60 was concluded after the NMR. [19] Adopted for the preparation of the flexible polymer electrolyte. Thin films solution-cast technique is. Where the PEO (Polymer) to NaI (Salt) (cation to monomer ratios was 0.065) were weighted separately and the different wt% of C60 is dispersed. Here the details about the film preparation can be understand from Saxena *et al.* (2014). [20]

3. MATERIAL CHARACTERISATION

The structural changes and the conductivity of the prepared films are observed. To find the structural changes, XRD was performed for all the samples. Further Impedance Spectroscopic (IS) techniques were used to measure conductivity. Impedance spectroscopy, was assessed in the frequency range of 100 Hz to 1 MHz. The cole-cole plot was drawn with the data collected through IS, and the bilk resistance (R_b) was calculated with curve fitting method. Finally, the conductivity for all the prepared samples were calculated and analysed graphically. [20]

4. RESULTS AND DISCUSSIONS

4.1 X-Ray Diffraction Study

To analyze the structural changes in fullerene dispersed PEO:NaI polymer thin flexible films, the XRD study was performed. In Figure 1 depicts the XRD plots for different polymer-salt matrix thin films dispersed with 3, 9, and 20 wt% of fullerene. In order to compare the result, the XRD plot for the pure polymer-salt thin films is also added to the graph.

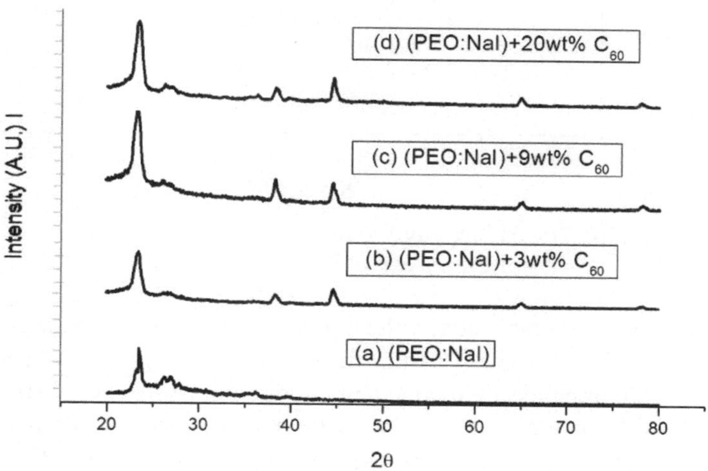

Fig. 1: The XRD Pattern of C_{60} Dispersed Polymer Electrolyte

During analysing the XRD graph, it can be seen that the peaks at 23 and 28 degrees are due to the PEO-NaI complex. here these peaks attribute to the partially crystalline and polymer-in-salt phase of PEO. [21] Also, it would be important to mention that there are no new peaks observed after the dispersion of C60. At the same time, the individual peaks at 38, 45, and 65 degrees in plots 1(b), 1(c), and 1(d) are seen. Here these peaks are attributed to the presence of C60 in his own phase, which indicates that no chemical reactions have taken place inside the system. Also, the increase in XRD peaks is due to the concentrations of C60. [22]

4.2 Electrical Conductivity Study

The conductivity of the different wt% of C60 dispersed polymer electrolyte thin films were measured with impedance spectroscopy data. Figure 2 shows the variation in conductivity for different thin films.

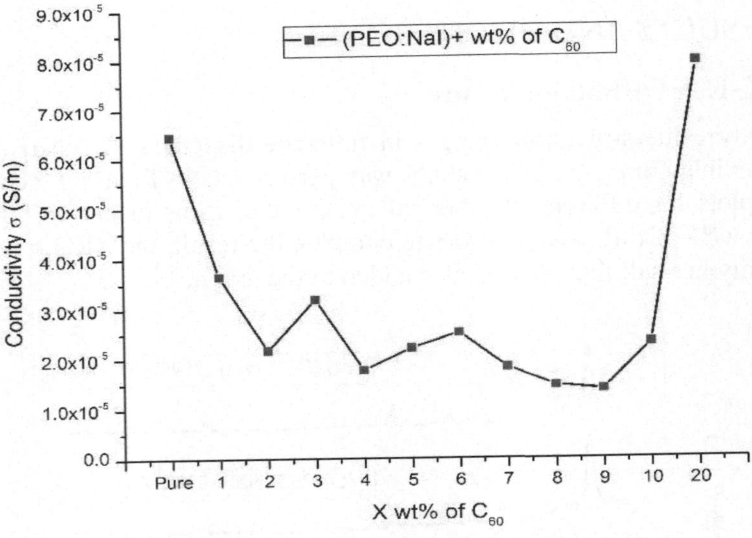

Fig. 2: Conductivity Variation for (PEO: NaI) + x wt% of C_{60} System at Room Temperature

Here it can be seen that, initially the conductivity decreases with the increase in the concentration of C60. But the further increase in the concentration of C60 (6 wt%) the appreciable increase in conductivity to 2.506×10^{-5} S/m can be seen and this increase in conductivity may be because of a change in crystallinity of the system. In the further process of dispersion of C60 onto the polymer-salt system, it was observed that the maximum conductivity of 7.920

$\times 10^{-5}$ S/m is at 20 wt% of C60. The random changes in conductivity (as shown in Figure 2) can be understood with the percolation model. [20]

5. CONCLUSION

The preparation of the flexible C60 dispersed polymer salt was done with the "Solution-Cast" technique. the maximum ionic conductivity was obtained for 20wt% of C60. the value of the highest conductivity was 7.95×10^{-5} S/cm. the composite behavior of the prepared samples was verified with XRD data. It was analysed that the flexible nature of the polymer electrolytes films with a good approximation of conductivity can be used for various e-textile applications. For further investigation, the samples are under process.

REFERENCES

[1] Suga T., Konishi H., Nishide H. (2007). "Photocrosslinked nitroxide polymer cathode-active materials for application in an organic-based paper battery". *Chemical Communications*, 1730–1732.

[2] Suga, T., Ohshiro, H., Sugita, S., Oyaizu, K. and Nishide, H. (2009). "Emerging N-Type Redox-Active Radical Polymer for a Totally Organic Polymer-Based Rechargeable Battery". *Advanced Materials,* 21, 1627–1630.

[3] Suga, T., Pu, Y.J., Kasatori, S. and Nishide, H. (2007). "Cathode- and Anode-Active Poly(nitroxylstyrene)s for Rechargeable Batteries: p- and n-Type Redox Switching via Substituent Effects". *Macromolecules,* 40(9), 3167–3173.

[4] Nystro¨m, G., Razaq, A., Strømme, M., Nyholm, L. and Mihranyan, A. (2009). "Ultrafast all-polymer paper-based batteries". *Nano Letters,* 9(10), 3635–3639.

[5] Tominaka, S., Nishizeko, H., Mizuno, J. and Osaka, T. (2009). "Bendable fuel cells: on-chip fuel cell on a flexible polymer substrate". *Energy & Environmental Science Journal, 2,* 1074–1077.

[6] Cai, W., Gong, X. and Cao, Y. (2010). "Polymer solar cells: Recent development and possible routes for improvement in the performance". *Solar Energy Materials & Solar Cells, 94,* 114–127.

[7] Lungenschmied, C., Dennler, G., Neugebauer, H., Sariciftci, S.N., Glatthaar, M., Meyer, T. and Meyer, A. (2007). "Flexible, long-lived, large-area, organic solar cells". *Solar Energy Materials & Solar Cells, 91,* 379–384.

[8] Dennler, G., Bereznev, S., Fichou, D., Holl, K., Ilic, D., Koeppe, R., Krebs, M., Labouret, A., Lungenschmied, C., Marchenko, A., Meissner D., Mellikov, E., Me´ot, J., Meyer, A., Meyer, T., Neugebauer, H., Pik, A.O., Sariciftci, N.S., Taillemite, S. and Hrle, T.W. (2007). "A self-rechargeable and flexible polymer solar battery". *Solar Energy, 81,* 947–957.

[9] Pushparaj, V.L., Shaijumon, M.M., Kumar, A., Murugesan, S., Ci, L., Vajtai, R., Linhardt, R.J., Nalamasu, O. and Ajayan, P. M. (2007). "Flexible energy storage devices based on nanocomposite paper". *PNAS, 104,* 13574–13577.

[10] Olivetti, E A., Kim, J.H., Sadoway, D.R., Asatekin, A. and Mayes, A.M. (2006). "Sol−Gel Synthesis of Vanadium Oxide within a Block Copolymer Matrix". *Chemistry of Materials,* 18(12), 2828–2833. https://doi.org/10.1021/cm060119s

[11] Lee, S.-H., Liu, P., Tracy, C.E. and Benson, D.K. (1999). "All- Solid- State Rocking Chair Lithium Battery on a Flexible Al Substrate". *Electrochemical and Solid-State Letters,* 2(9), 425.

[12] Pushparaj V.L., Shaijumon M.M., Kumar A., Murugesan S., Ci L., Vajtai R., Linhardt R.J., Nalamasu O. and Ajayan P.M. (2007). "Flexible energy storage devices based on nanocomposite paper". *PNAS*, 104(34), 13574–13577.

[13] Bhattacharya, R., de Kok, M.M. and Zhou, J. (2009). "Rechargeable electronic textile battery". *Applied Physics Letters*, 95, 223305.

[14] Kim, D.-H., Ahn, J.-H., Choi, W.M., Kim, H.-S., Kim, T.-H., Song, J., Huang, Y.Y., Liu, Z. and Rogers, J.A. (2008). "Stretchable and foldable silicon integrated circuits". *Science, 320*, 507–511. doi: 10.1126/science.1154367.

[15] Karaguzel, B., Merritt, C.R., Kang, T., Wilson, J.M., Nagle, H.T., Grant, E. and Pourdeyhimi, B. (2009). "Flexible, durable printed electrical circuits". *The Journal of The Textile Institute, 100*, 1–9.

[16] Hu, L., Choi, J.W., Yang, Y., Jeong, S., Mantia, F.L., Cui, L.-F. and Cui, Y. (2009). "Highly conductive paper for energy-storage devices". *PNAS*, 106(51), 21490–21494.

[17] Hu, L., Pasta, M., Mantia, F.L., Cui, L., Jeong, S., Deshazer, H.D., Choi, J. W., Han, S.M. and Cui Y. (2010). "Stretchable, Porous, and Conductive Energy Textiles". *Nano Letters*, 10(2), 708.

[18] Gao, Y., Xie, C. and Zheng, Z.J. (2021). "Textile Composite Electrodes for Flexible Batteries and Supercapacitors: Opportunities and Challenges". *Advanced Energy Materials*, 11, 2002838. https://doi.org/10.1002/aenm.202002838

[19] Singh, P.K., Bhattacharya, B. and Nagarale, R.K. (2010). "Effect of Nano-TiO2 Dispersion on PEO Polymer Electrolyte Property". *Journal of Applied Polymer Science*, 118(5), 2976–2980.

[20] Saxena, A., Singh, P.K. and Bhattacharya, B. (2014). "Structural and electrical studies of fullerene (C60) dispersed polymer electrolytes". *Materials and technology* (48)4, 485–490.

[21] Chatani, Y. and Okamura, S. (1987). "Crystal structure of poly(ethylene oxide)—sodium iodide complex". *Polymer*, 28 (11), 1815–1820.

[22] Ahmed, R.M. and El-Bashir, S.M. (2011). "Structure and Physical Properties of Polymer Composite Films Doped with Fullerene Nanoparticles". *International Journal of Photoenergy*, Article ID 801409, 6 pages. https://doi.org/10.1155/2011/801409.

Effect of Test Length on Okra Fiber Fineness

Pavan Kumar Gupta and Shamayita Patra*
Shri Vaishnav Institute of Textile Technology, SVVV, Indore, India
E-mail:*shamayitapatra@svvv.edu.in

ABSTRACT: *The fineness is very important property of a textile fiber. The fiber fineness decides the fineness of the manufactured yarn, processibility of a fiber in spinning and non-woven manufacturing, lustre of the fabric, handle of the fabric, etc. The measurement of the fiber fineness is crucial especially for the bast fibers. Bast fibers are multicellular in structure. Normally bast fibers are extracted through retting process. In the retting process, fiber bundles are not only separated from the bark of its plant stem but also within the fiber bundles. Due to this, fineness of the bast fibers is highly variable within the fiber length and between the fibers.*

In this study, fiber fineness of okra bast fiber was measured using three different test lengths (1.5 mm, 10 mm and 100 ± 20 mm). Variation in the fiber fineness was observed with the test lengths. Fineness of the okra fiber was found proportionate to the test lengths. This study is useful to understand the variation in fiber fineness within and between the natural bast fibers, especially, okra fibers.

Keywords: Okra Fiber, Water Retting, Fiber Fineness, Test Lengths, Optical Fiber Diameter Analyser.

1. INTRODUCTION

The fineness is a measure of the diameter or weight per unit length of fiber filament. [1] The fineness of the bast fiber is a genetic property that depends on plant age at harvest. [2] The fiber fineness for the natural bast fiber is very crucial in its grading process along with its length, strength, and color. [2] The fiber fineness is responsible for bending rigidity and torsional rigidity which ultimately affect the processing of fiber in spinning and non-woven manufacturing. [3,4]

The measurement of fiber fineness is very challenging, especially, for the natural bast fibers. [5] Generally, manmade fibers are manufactured with the same fineness throughout their length. It is presured that fiber fineness at any test length will be same. However, this is not true for the natural fibers such as cotton, wool, jute, etc. as natural fibers have high variation in their fineness within fiber (along its length) as well as between fibers. Further, the natural bast fiber added more complexity to the fiber fineness measurement as they have multicellular structure and extracted in bundle form from the bark of the plant's stem. Therefore, the fineness of the natural fiber, especially, the natural bast fiber may differ with the test lengths.

Like jute fiber, okra is also a natural lignocellulosic bast fiber, extracted from the bast of the okra plant stem through retting process. Okra (Abelmoschus esculentus) is a vegetable crop. It belongs to the Malvaceae family. It is mainly cultivated for its green tender fruits known as bhindi or lady finger. After fruit harvesting, the okra plant's harvest waste can be utilized to extract the technical bast fibers suitable for textile applications. [6–9]

To date, the extraction, and applications of okra fibers in textile applications are still in the research phase. Looking into its possible applications, the quality of okra fibers needs to be explored. Therefore, in this study, fiber fineness measurement of the okra bast fiber was carried out at three different test lengths (1.5 mm, 10 mm and 100 ± 20 mm). This study aims to understand the variation in fiber fineness within and between the natural bast fibers, especially, okra fibers.

2. MATERIALS AND METHODS

2.1 Materials

2.1.1 *Raw Material*

Okra fibers, extracted from the leftover okra plants harvest wastes, were taken as raw material in this study. Water was used as retting medium to extract the okra fibers.

Fig. 1: Okra Fiber Extracted Through Water Retting

2.1.2 *Instruments*

An Optical Fiber Diameter Analyzer (Make-Hornik, Model-OFDA 2000, Version 2203) was used to assess the diameter and fineness of fibers at 1.5 mm test length.

2.2 Methods

2.2.1 *Fiber Fineness Assessment Procedure*

In this study, fiber fineness was measured at three different test lengths to which authors defined as long length (100 ± 20 mm), medium length (10 mm) and short length (1.5 mm). The assessment procedure including sample preparation for okra fiber at different test lengths are mentioned below.

Fiber Fineness Assessment Procedure for Long Length (100 ± 20 mm)

Firstly, fibers were randomly collected which were of different lengths from the fiber strand. A total of 100 snippets were taken for analysis purpose for each sample. The selected fiber length varied from 40–180 mm with an average of 100 mm (SD ± 20 mm). Secondly, these 100 snippets of long cut length were measured by electronic weighing balance for weight in gram. Finally, the fiber fineness was calculated using the below formula.

$$\text{Tex} = \frac{\text{Weight of the fiber bundle (g)}}{\text{length of the fiber bundle (m)}} \times 1000 \quad \text{... (i)}$$

Fiber Fineness Assessment Procedure for Medium Length (10 mm)

Firstly, randomly taken fiber strands were cut to 10 mm length by a special cutter preciously. A total of 100 snippets (10 mm length) were taken for analysis purposes for each sample. Secondly, 100 snippets of 10 mm cut length were measured by electronic weighing balance for weight in gram while the fiber fineness was calculated using the formula (i).

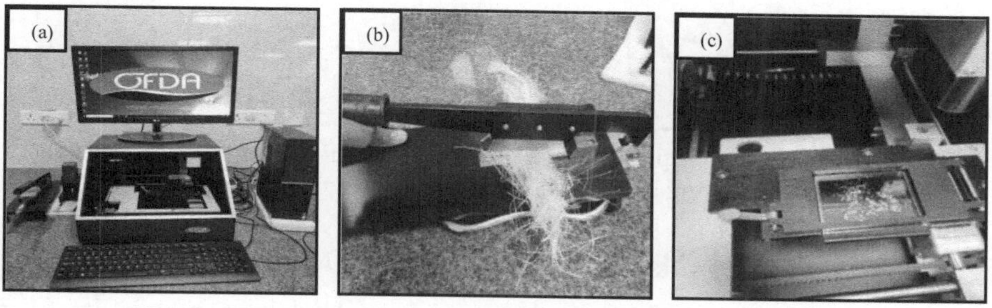

Fig. 2: a) OFDA Instrument, b) Cutter for 1.5 mm Cut Length, c) Sample Feed in Instrument for Micron Testing

Fiber Fineness Assessment Procedure for Short Length (1.5 mm)

To assess the fineness through diameter of the fiber, Optical Fiber Diameter Analyzer (Make-Hornik, Model- OFDA 2000, Version 2203) was used. Firstly, randomly taken fiber strands were cut to 1.5 cm length by a special cutter preciously. A total of 20000 snippets (1.5 mm length) were taken for analysis purposes for each sample (Refer Figure 2). Diameter of each snippet was measured optically, and then average diameter is converted into fineness using below formula. [10]

$$\text{Tex} = \left(\frac{\text{Fiber Diameter in micron}}{11.89}\right)^2 \times \frac{\text{Fiber Density}\left(\frac{\text{g}}{\text{cc}}\right)}{9} \qquad \ldots \text{(ii)}$$

Here, density of okra fiber was taken as 1.42 g/cc. [9]

2. RESULTS AND DISCUSSIONS

The fiber fineness of the bast fiber is a genetic property which depends upon the plant species. [2] Fiber fineness for natural bast fiber is very important in its grading process along with its length, strength, and color. Natural bast fiber are located in bundles in the bark of the plant stem and have multicellular structure (Refer Figure 3). [11] Therefore, fineness of the natural bast fiber may differ with the test length. To assess this phenomenon, the fineness of the okra fibers was measured at three different test lengths (100 mm, 10 mm, 1.5 mm) and was compared here.

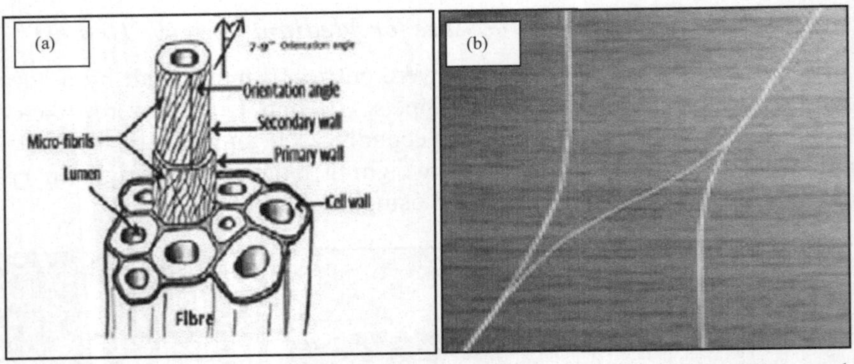

Fig. 3: (a) Schematic Representation of the Microstructure of a Jute Fibre, and (b) Branching of Fibers Due to Fiber in Bundle Form [2]

2.1 Comparison of Okra Fiber Fineness at Different Test Length

Okra fibersare extracted from the natural plant. High variation in its fiber fineness is natural. Therefore, 100 snippets were taken to average out the variation in individual fiber fineness measurement for long and medium test length while 20000 snippets for short test length in one reading. The average and standard deviation of 30 readings is represented graphically in Figure 4.

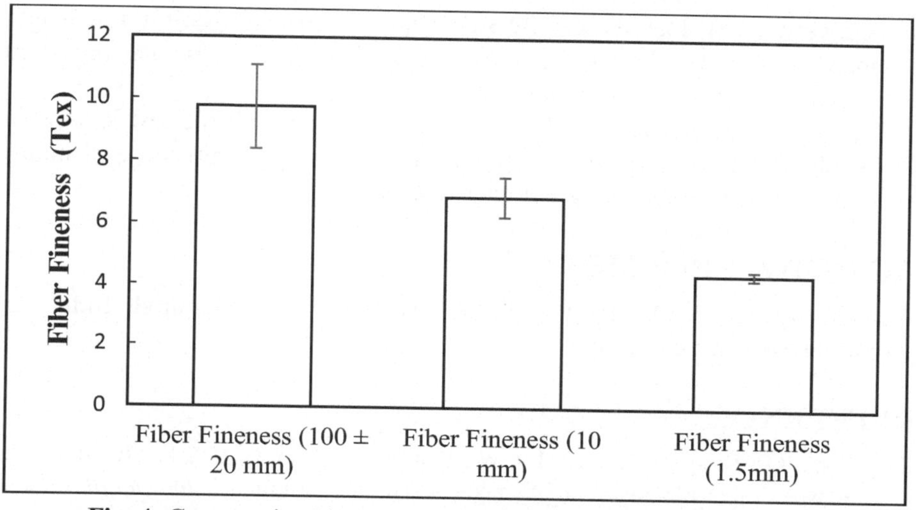

Fig. 4: Comparative Fineness of Okra Fiber at Different Test Length

Irrespective of retting conditions and okra plant lifespan, okra fibers showed distinct variation in the fineness at different test lengths. The resultant fineness of okra fibers at long, medium, and short test was 9.75 ± 1.35 tex, 6.89 ± 0.64 tex, and 4.39 ± 0.14 tex, respectively. Medium test length okra fiber showed 29.33% better fineness than at long test length while fineness at short test length still showed 36.28% better fineness than medium test length results.

Thus, lower test length showed better fineness than the longer test length. This may be due to following reasons. Longer the fiber test length, higher will be the probability of nodes due to branching or say meshy structure of fibers as shown in Figure 3(b). And there will be higher probability of impurities present over the fiber surface such as undissolved bark, foreign materials adhere to the fiber surface, etc. when the fiber test length is higher.

From this study, it can be stated that fineness of okra fiber varies with the test length and fiber fineness is proportionate to fiber length.

3. CONCLUSION

Fineness of the bast fibers is highly variable within the fiber length and between the fibers as bast fibers are multicellular in structure. Measurement of the fiber fineness is very crucial for any bast fibers. In this study, fiber fineness of the okra bast fiber was measured using three different test lengths (100 ± 20 mm, 10 mm and 1.5 mm).

Irrespective of retting duration and okra plant lifespan, okra fiber fineness showed distinct variation in the fineness at different test lengths. Fineness of okra fiber found to be 9.75 ± 1.35 tex, 6.89 ± 0.64 tex and 4.39 ± 0.14 tex respectively. Okra fiber showed 29.33% better fineness at medium test length compared to long test length. Similarly, 36.28% better fineness was observed for shorter test length over medium test length.

Thus, variation in the fiber fineness was observed with the test lengths in this study which may be due the higher probability of the presence of nodes and external impurities with the higher test lengths.

ACKNOWLEDGEMENT

Authors thanks to Mr. Pankaj, Sintex Textile Private Limited, India for providing the testing facility.

REFERENCES

[1] Geng, Q., Zhou, C., Nie, K., Lv, W., Ben, H. and Han, G. (2021). "Relationship between Fiber Fineness and Diameter of Three Bast Fibers". *Journal of Natural Fibers* 2021.doi:.10.1080/15440478.2021.1877233.

[2] Ammayappan, L., Samanta. K.K. and Ghosh, R.K. (2017). "Training manual on recent advances in processing technologies for value addition of jute and allied fibers". Ammayappan, L., Samanta, K.K., Ghosh, R.K. editors. ICAR-National Institute of Research on Jute and Allied Fiber Technology, Kolkata, 1–189.

[3] Korkmaz, Y.A. and Behery, H.M. (2004). "Relationship Between Fiber Fineness, Break Draft, and Drafting Force in Roller Drafting". *Textile Research Journal*, 74(5), 405–8. doi: 10.1177/004051750407400506.

[4] Dhange, V.K., Kadole, P.V. (2019). "Nonwoven performance from fashion apparel perspective". *Man-Made Text India*, 47(9), 298–301.

[5] Grimes, M.A. (1942). "Measuring Fiber Fineness". *Textile Research Journal*, 13(1), 12–8. doi: 10.1177/004051754201300104.

[6] Gupta, P.K. and Patra, S. (2019). "Effect of retting period on okra fiber optical property". Dhar, S., Dhar, U., Sainy, M., Mishra, S.D. editors. *Innovation Invention and Industrialization emerging issues*. Excel India Publishers. 43–52.

[7] Alam, M.S. and Khan, G.M.A. (2007). "Chemical analysis of okra bast fiber (Abelmoschus Esculentus) and its physico-chemical properties". *Journal of Textle Apparel*, Technol Manag. 5(4), 1–9.

[8] Tripathi, K.K., Warrier, R., Govila, O.P. and Ahuja, V. (2011). "Biology of Abelmoschus esculentus L. (Okra). *Ministry of Environment and Forests*, Government of India, New Delhi, Available from: https://biosafety.icar.gov.in/wpcontent/uploads/2015/11/Biology_of_Okra1.pdf

[9] Gupta, P.K., Patra, S. and Samanta, K.K. (2021). "Potential of Okra for Application in Textiles: A Review". Vol. 18, *Journal of Natural Fibers*, 1788–800.

[10] Minifibers, I. Choosing The Proper Short Cut Fiber for Your Nonwoven Web. Available from: https://docplayer.net/20704274-Choosing-the-proper-short-cut-fiber-for-your-nonwoven-web.html

[11] Wang, H.M., Wang, X. (2004). "Evaluation of the fineness of degummed bast fibers". Fibers Polym, 5(3), 171–6. doi: 10.1007/BF02902995.

Effect of Dyeing on Properties of Polyester/Viscose Blended Yarns

Mandar S. Kulkarni* and U.J. Patil
D.K.T.E Society's Textile and Engineering Institute, Ichalkaranji
E-mail: *maandar13@gmail.com

ABSTRACT: *The change in yarn properties of Ne30/1 Polyester/Viscose blended yarn spun with different blend proportions and by different spinning technology after dyeing with different depth of shade have been studied. The yarn properties of the grey yarn have also been tested for comparison. Percent change in yarn properties before and after dyeing were statistically evaluated. During dyeing the polyester component is first dyed with disperse dye followed by the dyeing of viscose component with reactive dye. After dyeing, finishing is done with a softener. It is observed that, there is significant change in yarn properties like strength, hairiness, yarn to metal friction etc. Amongst the three parameters under study: Blend Proportion, Yarn Spinning Technology and Depth of shade; yarn spinning technology has more effect on change in yarn properties after dyeing irrespective of blend proportion and depth of shade followed by blend proportion and depth of shade has least influence on change in yarn properties.*

Keywords: Blended Yarn, Spinning Technology, Polyester/Viscose Blend, Depth of Shade.

1. INTRODUCTION

The dyed yarns are used in making stripe knit/woven fabrics, solid dyed yarn fabric or in sweater manufacturing. In the yarn dyeing process yarns are dyed in package form or hank form. Depending on the individual procedure of a textile engineer or a dyeing master dyeing process may vary. But the typical flow chart of yarn dyeing remains the same. Yarn dyeing is done either to improve aesthetic or functional value of yarn. But in order to understand the dyeing procedure, it is necessary to know the type of yarns and their blends used. In this case, polyester and viscose yarns in different blend ratio spun from different spinning technology are used. When two varied materials are blended to form a new product, their best qualities get mixed and enhanced as well. For example: polyester provides strength, viscose gives softness, lightness, comfort and skin-friendliness. Present study deals with manufacture and study of Polyester/Viscose blend yarn with different blend proportions spun on single and double nozzle air jet spinning in comparison with conventional ring spinning technology. [1-7]

2. MATERIALS AND METHODS

2.1 Material

In this study, polyester/viscose blended yarn is used as a raw material. Polyester & viscose fiber have 1.2 denier and 38mm length. Yarns having linear density of Ne30 were spun from three spinning technologies i.e. Ring, Single Nozzle AirJet and Double NozzleAirJet.

2.2 Method

The Taguchi design of the experiment is used to spin the yarn. Same method is continued while dyeing the yarns with different depths. The detailed design of experiment is shown in Table 1 below.

Table 1: Design of Experiment

Sr. No.	*Blend Proportion (Polyester/Viscose)*	*Technology of Spinning*	*Depth of Shade*
1.	50/50	RING SPUN	LIGHT
2.	50/50	DOUBLE NOZZLE VORTEX	MEDIUM
3.	50/50	SINGLE NOZZLE VORTEX	DARK
4.	67/33	RING SPUN	MEDIUM
5.	67/33	DOUBLE NOZZLE VORTEX	DARK
6.	67/33	SINGLE NOZZLE VORTEX	LIGHT
7.	80/20	RING SPUN	DARK
8.	80/20	DOUBLE NOZZLE VORTEX	LIGHT
9.	80/20	SINGLE NOZZLE VORTEX	MEDIUM

2.2.1 *Dyeing Procedure*

The dyeing process was carried out according to the conventional two-step dyeing method. The polyester part of the blended yarn was first dyed with disperse dye at 130° ± 1°C followed by reduction clearing and washing. In the second step, viscose was dyed with reactive dyes at 80° ± 1°C. Finishing of the yarn was done with softener. Navy blue colour dye is used with 0.5%, 1.0% and 1.5% depth of shade. After dyeing soft packages were converted to hard packages to carry out the yarn testing. To investigate the effects of parameters under study on the yarn properties, t-tests were performed at 5% (0.05) levels of significance.

2.2.2 *Testing*

Yarn properties before and after dyeing are tested and evaluated for percent change with respect to gray yarn.

3. RESULTS AND DISCUSSIONS

3.1 Effect of Parameters under Study on Percent Change in Yarn Count

Figure 1 shows the effect of parameters under study on percent change in yarn count. Yarn count at grey stage for all the samples was maintained within the tolerance limit by proper selection of draft. After dyeing, the count of all the samples has become coarser. None of the parameters under study has a significant influence on change in yarn count. Change in yarn count after dyeing, irrespective of depth of shade is statistically insignificant.

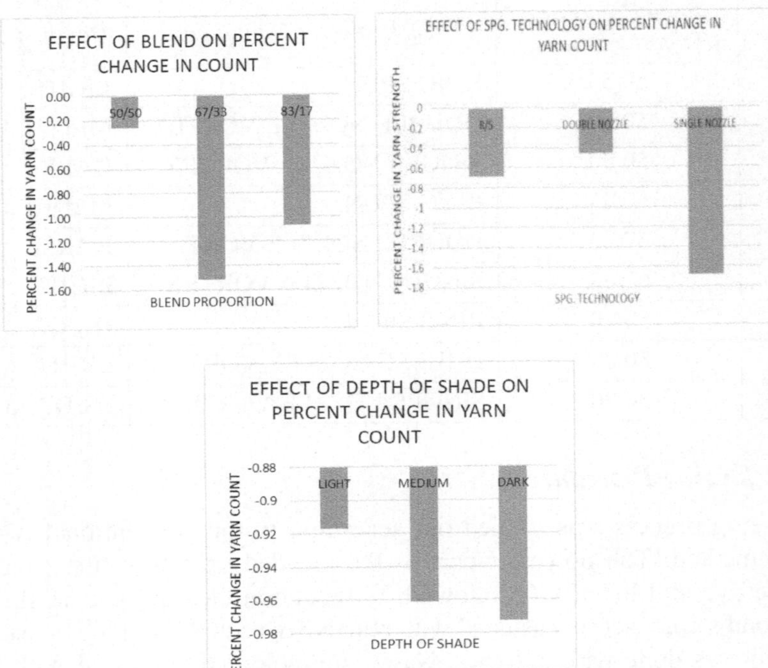

Fig. 1: Effect of Parameters on Yarn Count

3.2 Effect of Parameters under Study on Percent Change in Yarn Diameter

Figure 2 shows the effect of parameters under study on percent change in yarn diameter. From the graphs, it is seen that after dyeing, there is significant increase in yarn diameter. This increase may be attributed to swelling of viscose component during dyeing. Levels of parameters under study seem to have no significant influence on percent change except depth of shade.

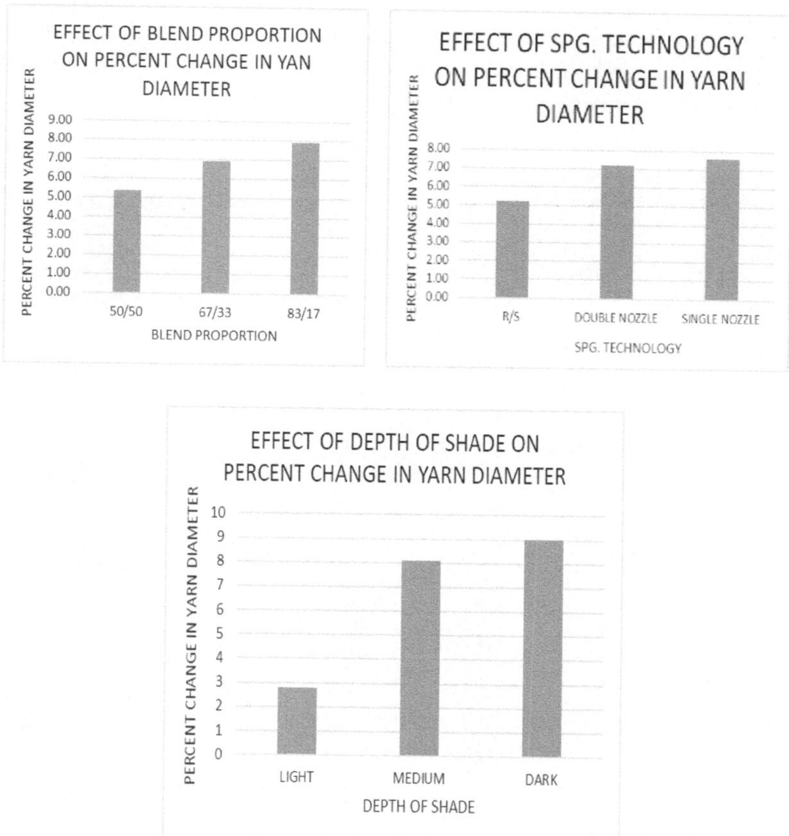

Fig. 2: Effect of Parameters on Yarn Diameter

3.3 Effect of Parameters under Study on Percent Change in Single Yarn Strength

From the above Figure 3, it is observed that, incase of blend ratio 50/50 and 67/33, there is no significant drop in yarn tenacity. But in case of blend ratio 83/17, it is observed that the drop in tenacity is significant, this may be because after the dyeing processing to remove the oligomers and unfixed dye from the surface, the post dyeing treatment i.e. reduction clearing is carried out. This process is carried out by using sodium hydroxide (1 gpl) and sodium hydroshulphite (1 gpl) at 80°C for 10 mins. This process is carried out 2 times to havebetter effect. As the sodium hydroxide is used, under the right conditions, the hydrophobicester groups in polyester fibers can be hydrolyzed to form hydrophilic groups, i.e. carboxylic and hydroxyl groups. Polyester undergoes nucleophilic substitution and is hydrolyzed by aqueous sodium hydroxide. The hydroxyl ions attack the electron-deficient carbonyl carbons of the polyester to form an intermediate anion. Chain scission follows and results in the production of hydroxyl and carboxyl ate end-groups. Hence the polymeric chains get broken, resulting in drop in tenacity.

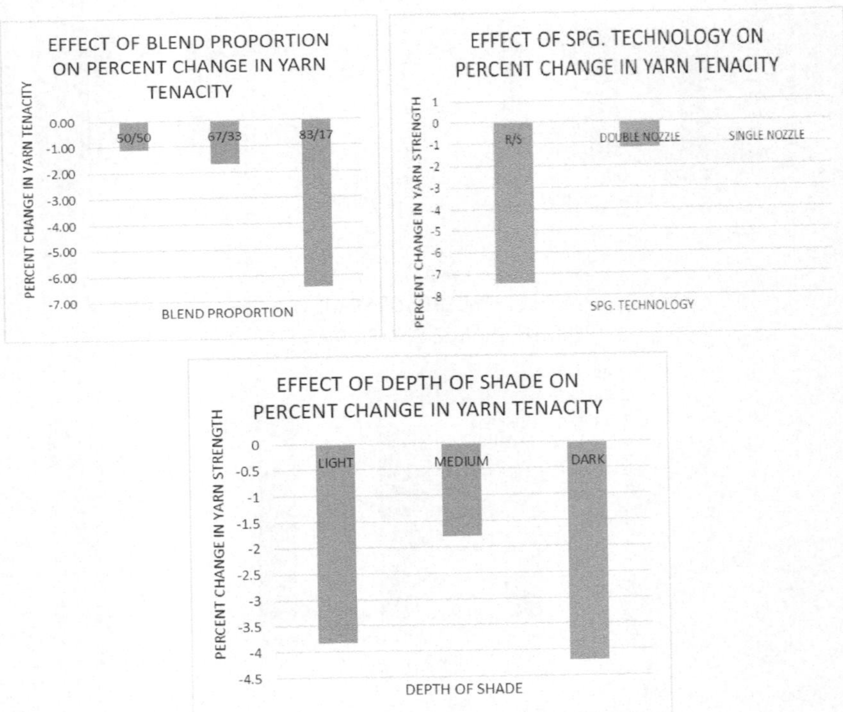

Fig. 3: Effect of Parameters on Yarn Tenacity

3.4 Effect of Parameters under Study on Percent Change in Single Yarn Elongation

Figure 4 shows the effect of parameters under study on percent change in yarn elongation. From above graphs, it is observed that there is significant increase in yarn elongation after dyeing. Blend proportion does not show any trend with respect to percent increase. Percent increase in elongation after dyeing is more in case of vortex yarn against ring spun yarn, especially single nozzle vortex yarn shows more increase.

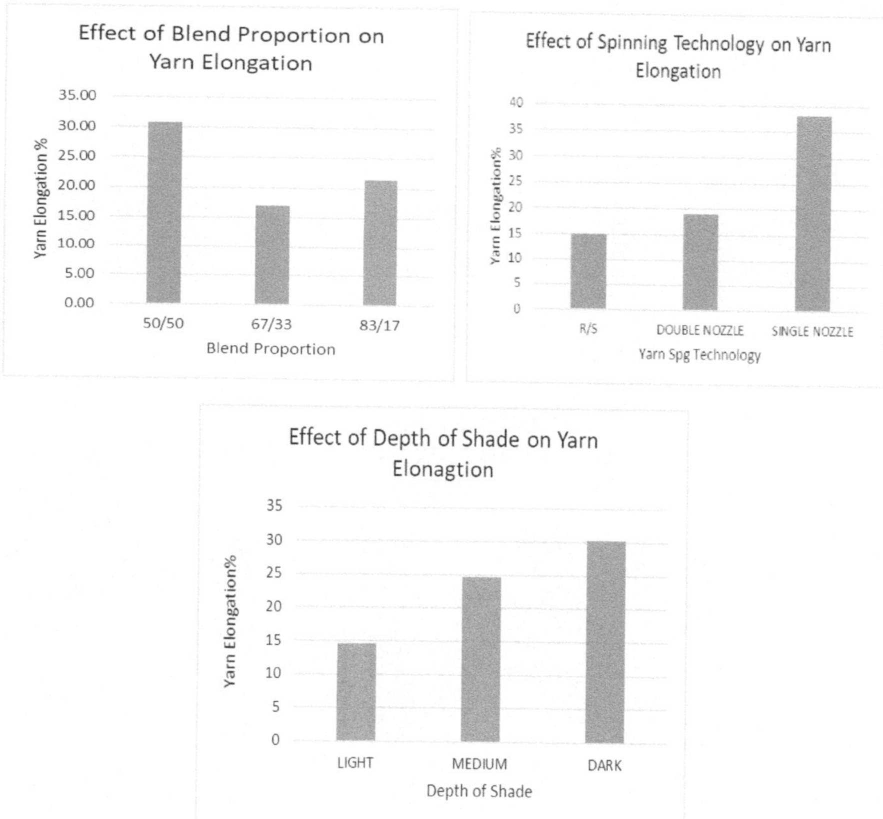

Fig. 4: Effect of Parameters on Yarn Elongation

3.5 Effect of Parameters under Study on Percent Change in Yarn to Metal Friction

Figure 5 shows the effect of parameters under study on percent change in yarn to metal friction after dyeing. Friction in a fibrous assembly mainly depend on the characteristics of the surfaces which are mutually in contact, in addition to the magnitude of the normal pressure acting between them. Significant drop in yarn to metal friction after dyeing has been observed in all the samples. Drop in friction is attributed to the softening process followed after dyeing. Drop in friction values is less in the yarns wherein proportion of polyester fibre is more.

As far as spinning technology is concerned, drop is more with vortex yarn in comparison with ring yarn. This may be due to less hairiness of vortex yarn. Drop in S3 value has been noticed after dyeing, irrespective of depth of shade. Softening treatment might be making yarn more circular and uniform as far as surface is concerned, resulting into low yarn to metal friction after dyeing.

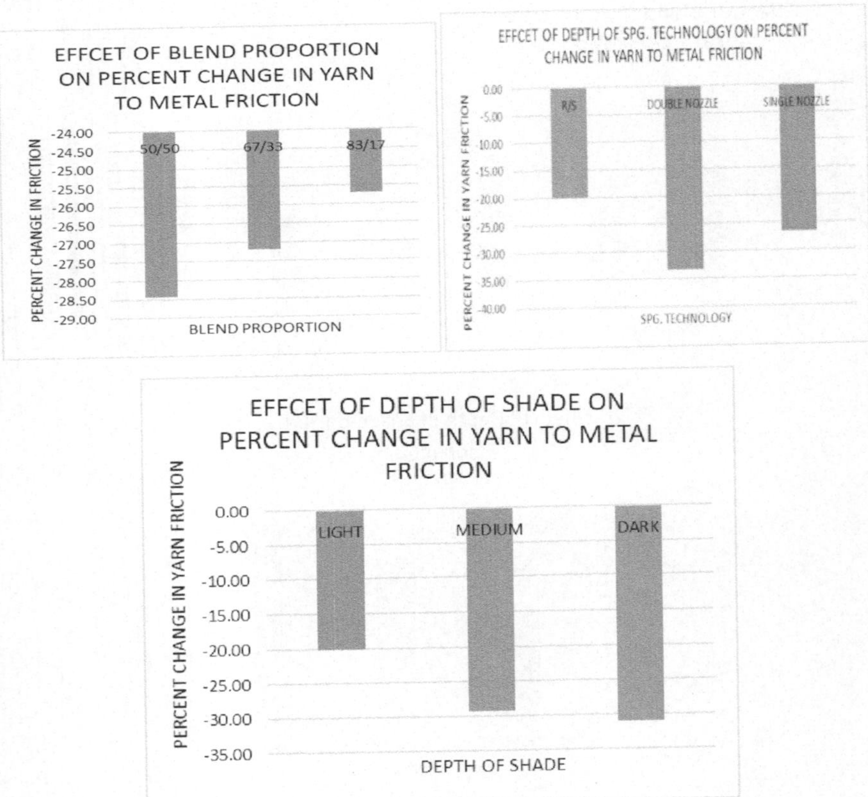

Fig. 5: Effect of Parameters on Yarn to Metal Friction

3.6 Effect of Parameters under Study on Percent Change in Yarn Unevenness

Figure 6 illustrates the effect of parameters under study on percent change in yarn unevenness after dyeing. From results, it is seen that there is no significant change in yarn unevenness. But, with regard to yarn spinning technology, it is observed that yarn spun by a double nozzle vortex system shows an increase in yarn unevenness after dyeing whereas, yarn spun from a single nozzle shows no increase after dyeing. Some more studies with respect to change in yarn unevenness of vortex yarn are required to confirm the result.

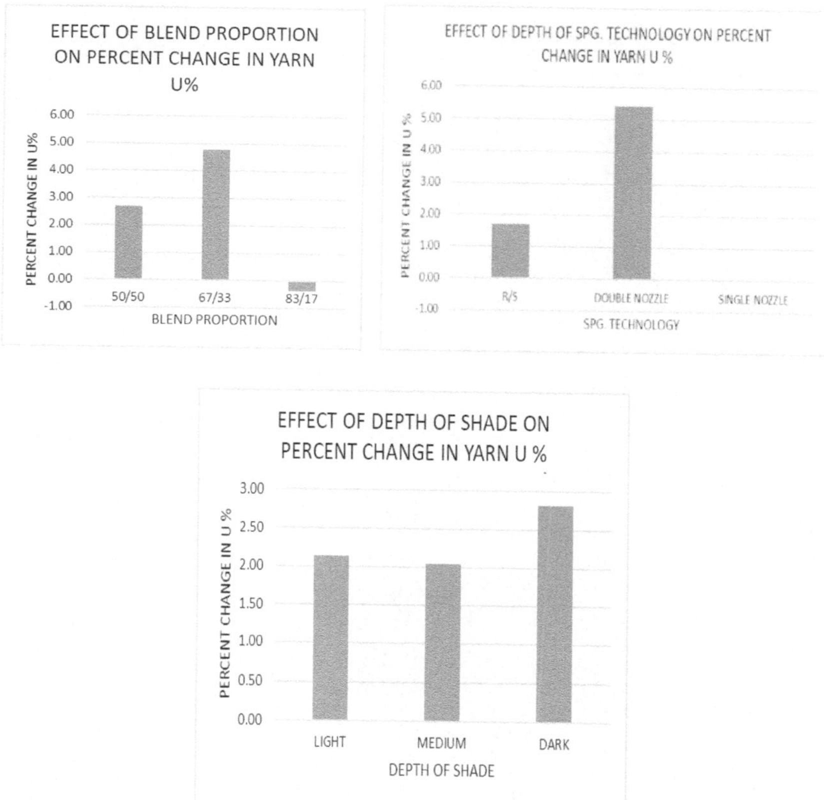

Fig. 6: Effect of Parameters on Yarn Unevenness%

3.7 Effect of Parameters under Study on Percent Change in Yarn Imperfection

Imperfections after dyeing have reduced irrespective of blend proportion. The difference in imperfections after dyeing for different blend proportions is statistically insignificant. Imperfections of Ring and Double Nozzle Vortex yarn have increased by around 10 percent after dyeing. But significant reduction in imperfections of single nozzle vortex yarns is observed after dyeing, mainly in terms of thick places and neps. Yarn structure seems to have a significant influence on yarn imperfections. Depth of shade does not show any trend as far as increase or decrease in imperfections are concerned. As the evenness tester works with capacitance principle, dielectric constant of the yarn after dyeing with different depth of shade might be getting changed affecting the sensitivity towards thick places and neps, which need to be confirmed.

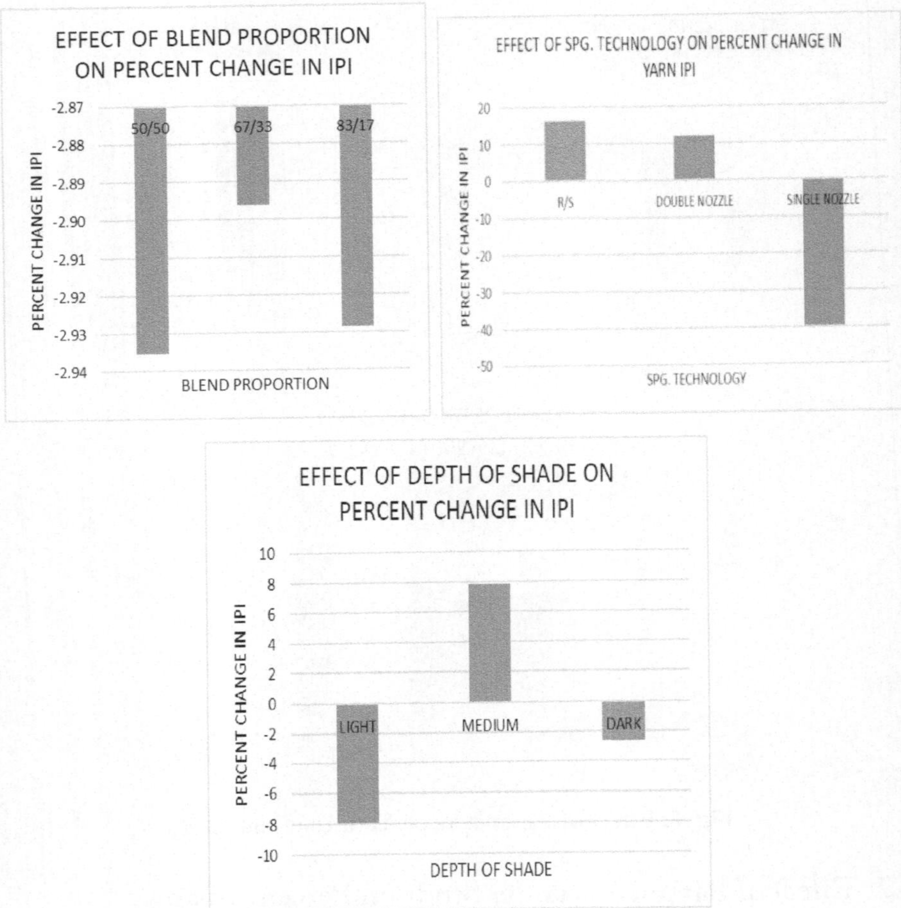

Fig. 7: Effect of Parameters on Yarn Imperfections/Km

3.8 Effect of Parameters under Study on Percent Change in Yarn Hairiness

Yarn hairiness on UT 5 works with light scattering principle. In this case, the yarn body is treated as a dark body and fibres protruding out of the dark body will scatter the light depending upon the length of protrusion. In case of dyed yarn, yarn body as well as protruding fibres both will be observed dark. Therefore, generally it has been observed that hairiness index after dyeing reduces. Therefore, it is advisable to test yarn hairiness in terms of total number of protruding fibres more than standard length for fixed measured length (Like S3 value measured on Zweigle Hairiness Tester).

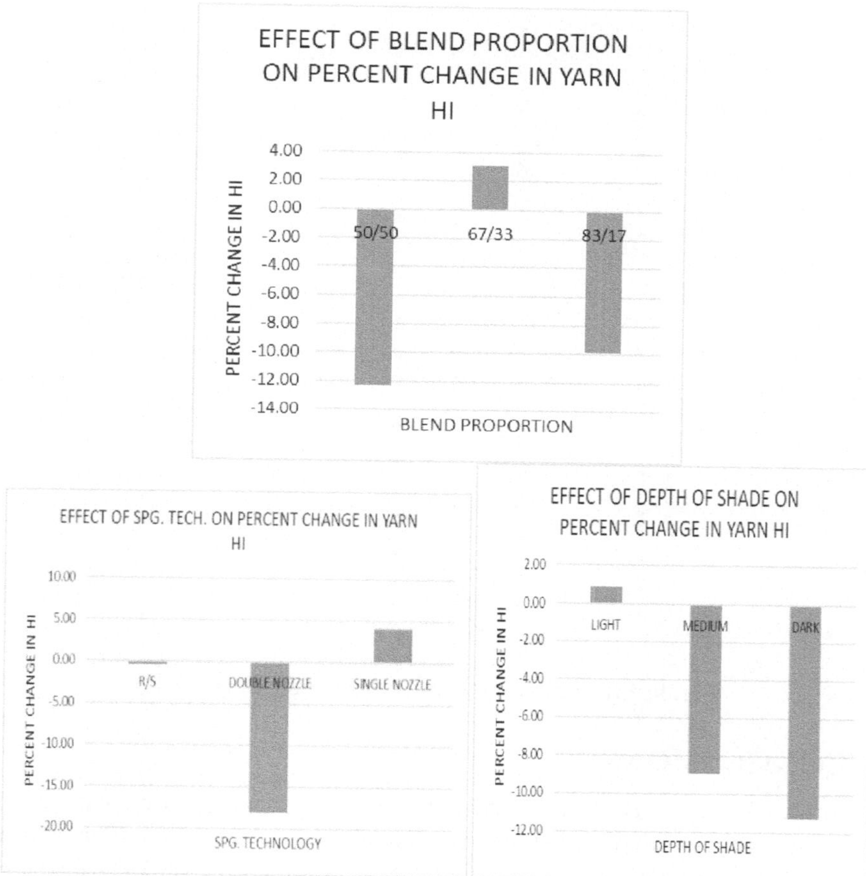

Fig. 8: Effect of Parameters on Yarn Hairiness Index

3.9 Yarn Hairiness S3 Value

It is observed that S3 value reduces after dyeing. Percent reduction in S3 value for different blend proportions does not vary significantly. The S3 value of all the yarns reduces significantly. But percent reduction is more for the yarn spun with single nozzle vortex technology. After dyeing, softening treatment is given to all the yarns. During this process, fibres might have been compacted and yarn structure becoming more of solidaire. The results match with the results of yarn to metal friction confirming the phenomenon.

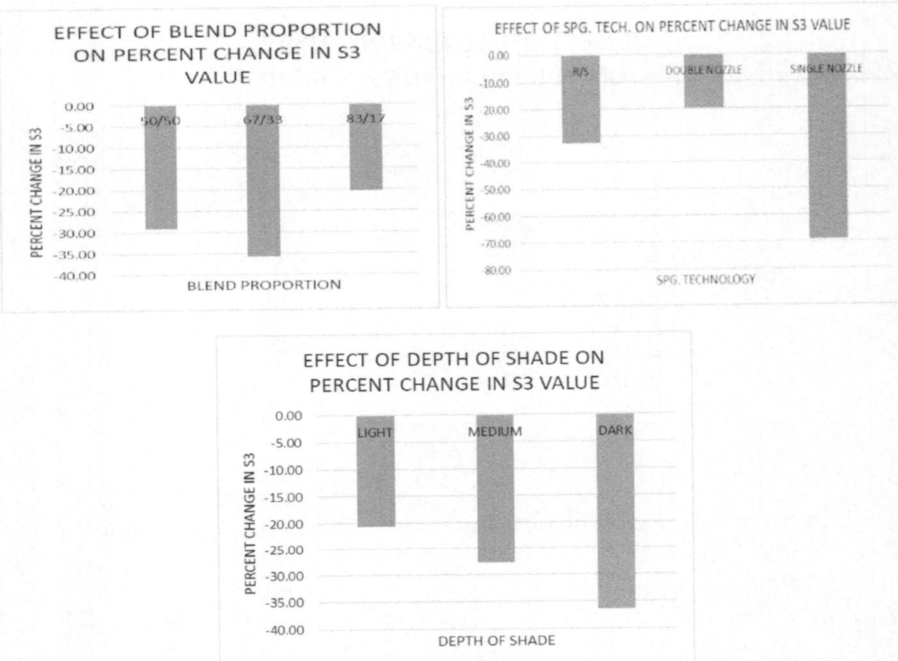

Fig. 9: Effect of Parameters on Yarn Hair Count (S3)

4. CONCLUSION

From the above study it has been observed that the yarn properties except yarn unevenness and imperfections significantly change after dyeing, irrespective of depth of shade. Amongst the parameters under study, yarn spinning technology is found to have most influence followed by blend proportion and depth of shade has least influence. Hence, it can be concluded that yarn structure and fibre component in the yarn are the influencing factors for percent change in yarn properties after dyeing.

REFERENCES

[1] Sengupta, A.K., Bajaj P. and Sen K. (1983). "Simultaneous Dyeing and Texturing of Polyester/Viscose Blended Yarn". *Textile Research Journal*, 53(4):258–263. doi: 10.1177/004051758305300409

[2] Memon, H., Khoso, N.A. and Memon S. (2015). "Effect of dyeing parameters on physical properties of fibers and yarns". *Journal of Applied Sciences and Engineering Research*, 4(4): 401–407.

[3] Shaw, V.P. (2019). "Effect of dyeing process on tensile strength properties of polyester/cotton blended yarns". *Indian Journal of Fibre & Textile Research*, 44, 369–372.

[4] Tyagi, G.K., Ajay, R., Gupta, S. and Kapoor, V. (1986). "Influence of Fibre Linear Density, Twist and Direct Dye on Physical and De-uptake Characteristics of Viscose Rayon Yarns". *Indian Journal of Textile*, 11, 150–155.

[5] Maruf, N.H. and Haque, M.M. (2016). "Evaluation of Processing Performance and Properties of 100% Cotton and Cotton-Polyester Blended Ring-Spun Yarn". *International Journal of Current Engineering and Technology*, 6(2). DOI: 10.4172/2165-8064.1000244

[6] Manik, P., Majumder, S. and Hossain, R.K. (2019). "Effect of Blend Ratio on Quality Characteristics of Polyester/Cotton Blended Ring Spun Yarn". *Trends in Textile Engineering & Fashion Technology*, 5(1). DOI: 10.31031/TTEFT.2019.05.000601.

[7] Erdumlu, N., Ozipek, B., Oztuna, A.S. and Cetinkaya, S. (2009). "Investigation of Vortex Spun Yarn Properties in Comparison with Conventional Ring and Open-end Rotor Spun Yarns". *Textile Research Journal*, 79(9):585–595.

A Novel Bio-Composite for Removal of Heavy Metals from Waste Water by Adsorption Technique

Subhankar Maity

Uttar Pradesh Textile Technology Institute, Kanpur, Uttar Pradesh, India
E-mail: smaity@uptti.ac.in

ABSTRACT: *In the advent of industrialization and human activities, the discharge of waste and wastewater containing heavy metals in environment is increasing day by day and becoming a threat to vegetation and mankind. Various industries such as leather, textile, paper mills, fertilizer, pesticide, metallurgy, etc. discharge liquid wastes containing different heavy metals into the environment. They can be absorbed and accumulated in human body and caused serious health effects like cancer, organ damage, nervous system damage, and in extreme cases, death. Also, it reduces growth and development of natural vegetation and all other living beings. Allowable concentration of Cr, Ni, Cu, and Zn are 0.5, 0.2, 1 and 5 mg/L respectively as per WHO guidelines for drinking water quality (2014). In recent years, various methods of removal of heavy metal from wastewater have been extensively studied. Most promisingly, the conducting polymers like polypyrrole (PPy), polyaniline (PAni), etc. are used for treating waste water due to their inherent properties. This research has been focused on the development of PPy coated sorbents for the removal of heavy metals from industrial waste water.*

PPy coated adsorbents had good adsorption capacities for Mg, Fe, Cu, Cd, Pb, Zn and Ni. PPy coated saw dust could be able to remove Cr from waste water. This method would be a replacement for costly conventional methods of removing heavy metal ions from wastewater. It is expected that this method would be an alternative one for waste water treatment which will benefit the industries in future.

Keywords: Polypyrrole, Jute, Adsorption, Wastewater Treatment, Heavy Metal.

1. INTRODUCTION

Contamination of water with heavy metals is a new form of water pollution and becoming a serious threat to all living beings day by day.

If the density of any metal exceeds 5 gcm^{-3}, considered as heavy metals. [1] However, any toxic metal can be called as heavy metal irrespective of its density or atomic number. Various industries like mining, leather, textiles, paper mills, fertilizer, pesticide, metallurgy, surface finishing, etc. generate and discharge waste water into the environment that contained various heavy metals. There are three kinds of heavy metals contaminating water viz.

(i) Toxic metals such as (Cr, Pb, Hg, Cd, As, Ni, Co, Zn, Cu, Sn, etc.)
(ii) Precious metals such as (Au, Ag, Pt, Pd, Ru etc.) and
(iii) Radio nuclides such as (U, Th, Am, Ra, etc.).

After consuming the contaminated water, the metals are accumulated in human body and caused serious health effects like damage to vital organs, nervous system failure, hormonal imbalance, and in extreme cases cancer. Also, it impairs growth and development of natural vegetation and all other living beings. Thus, metals as contamination bring serious environmental pollution, threatening human health and ecosystem when they are consumed beyond a tolerance limit. [2] Some metals are carcinogenic in nature which highly damage our body. If the waste water discharged directly into the natural water it causes high risk for the aquatic eco system. The direct discharge into the soil highly ruined the fertility of soil. The metals which are mainly present in waste water are Zn, Cr, Fe, Ni, Cu, Pb, Cd and Mg. These metal-contained wastewater are directly or indirectly discharged into streams, lakes, rivers or oceans, especially in developing countries. [3–4] Moreover, the soils surrounding military bases offer a potential risk of certain heavy metals, which can be a threat to groundwater and surface water. The presence of chromium, lead, and cadmium are cited as the most often accruing soil contaminants in military base. These are due to practices in military such as electro-plating, mechanical work, aircraft rebuilding, battery acid disposal, leaking tanks, firing ranges, pistol ranges, open burning of ordnance, detonation, etc. These heavy metal ions are not bio-degradable in nature and when consumed are easily accumulated in living organisms. [2] Although consuming a limiting amounts of different heavy metals is extremely necessary for human beings for the synthesis of vitamins or utilizing as co-factors, excessive exposure or intake can have dangerous consequences. Such as Zinc (Zn) is an essential trace element good for human health. However, excessive Zn can cause serious health issues like stomach cramps, vomiting, nausea, skin irritations, and anemia. [4–5] Copper (Cu) performs vital work for human metabolism. But the excessive consumption of copper is leading to serious toxicological issues like vomiting, convulsions, cramps, and even death. [8] Chromium(Cr) is available in the aquatic environment mainly in two states: Cr(III) and Cr(VI). In general, Cr(VI) is more toxic than Cr(III) and it can easily penetrate through cell membrane. It accumulates in food chain and consumed by animals and human beings naturally. It therefore affects animal and human physiology and causes severe health issues like skin irritation, lung carcinoma etc. [5-6] Wastewater regulations are established in some countries to reduce human and environmental exposure to hazardous chemicals. In the same line, the Central Pollution Control Board (CPCB, India), set various standard for the discharge of waste water effluent into any water body, for example BOD 30 ppm, COD 250 ppm, pH 5.5–9.0, Manganese (mg/l): 1.0, Chromium (mg/l): 0.05 etc. [1] Therefore, the removal of traces of heavy metals from contaminated water has been a special concern for researchers and environmental scientists. Conductive polymers such as polypyrrole (PPy) coated sorbents can be the potential materials for the removal of chromium ions from contaminated water. Use of these conductive polymers for this purpose is gaining popularity day by day due to their ease of synthesis, biocompatibility, redox and ion exchange properties. This research is therefore focused on the physico-chemical properties, adsorption characteristics, mechanism and efficiency of different polypyrrole-based adsorbents for chromium removal from contaminated water.

2. MATERIALS AND METHODS

2.1 Materials

Saw dust was collected from wood furniture shop in Kanpur, (U.P. India), Jute fabric (250 g/m^2 (EPI 22, PPI 23) was collected from local Shop in Kanpur, U.P. India.

2.2 Chemicals

Pyrrole is purchased from Sigma Aldrich, India. $FeCl_3$, Sodium hydroxide, and Sulphuric Acid were purchased from CDH, India. All chemicals are laboratory grade and used as received.

2.3 Preparation of polypyrrole Coated Wood Saw Dust (SD) Jute Bio-Composite for Chromium Removal

Sawdust (SD) is an abundant by-product of wood industry and easily available at low cost. It is primarily a cellulosic material contains various organic compounds with polyphenolic groups. Therefore, it can be easily coated with PPy to obtain PPy/SD composites, which can adsorb heavy metal ions from water. [7–8] Preparation of PPy coated saw dust and jute is demonstrated in Figure 1. First, oxidant and monomer solutions are prepared separately by adding 5 g $FeCl_3$ and 1 mL pure pyrrole to 100 mL of deionized water in two separate beakers and stirring by a magnetic stirrer. Then, 1 g of SD is added to oxidant solution and monomer solution is added. The mixture is kept on stirring for 4 hr at room temperature. As a result, oxidative polymerization of pyrrole commences to form polypyrrole which is coated on the surface of saw dust and jute fibre. Finally, the PPy coated saw dusts and jute fabric prepared. The coated SD are filtered from the solution and jute fabric is simply taken out from the solution and washed several times with deionized water and dried at room temperature.

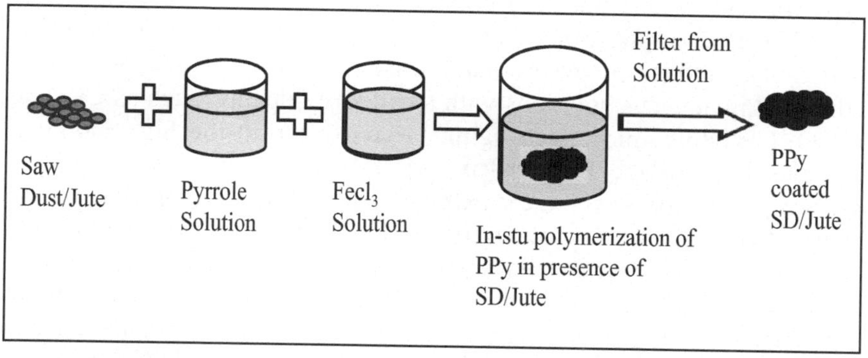

Fig. 1: Preparation of Polypyrrole Coated Wood Saw Dust (SD) and Jute Bio-Composites

2.4 Cr Removal Technique

Adsorption process can be successfully adopted to remove heavy metals from the wastewater. Adsorption is the process where free molecules are forced to accumulate on the surface of the sorbent. The suspended molecules which are present in bulk form in a solution are being adsorbed into the pores of a suitable adsorbent. The driving force for adsorption is the ratio of the concentration to the solubility of the compound. Adsorption can be divided into two groups which are Physical Sorption and Chemisorption. Physical adsorption creates weak intermolecular forces while chemisorption forms essential chemical bond between the sorbent molecule and the surface of the adsorbent. On the view of the demerits of other waste water treatment processes, this adsorption process may be a suitable alternative process which would be efficient and economic. This adsorption process suitable for heavy metal removal in a more effective way as compare to other processes. [9]

- 0.5 g of PPy coated Jute/saw dust sorbents agitated with 50 mL 10 ppm Cr solution at 30 ± 0.5°C for 25 min.
- At the end of predetermined time intervals, the sorbent was filtered and the reduction of concentration of Cr, was determined by MPAES (Microwave Plasma Atomic Emission Spectrometer).

- 25 minutes of adsorption experiments were performed by agitating 1.0 g of sorbent with 100 mL Cr (10 ppm) solution of at 30 ± 0.5°C. The detail treatment process is shown in Figure 2.

3. RESULTS AND DISCUSSIONS

3.1 Polypyrrole as a Material of Ion Absorber

Polypyrrole (PPy), a conducting polymer has been extensively explored for various potential applications due to its good environmental stability, high conductivity, redox properties and ease of process ability. [3] PPy synthesized in aqueous solutions with small size dopants, such as Cl^-, ClO_4^-, and NO_3^-, exhibits anion exchanging behavior due to the high mobility of these ions through the PPy matrix [9]. PPy intercalated with Cl^- ion is possessing great ion exchange capability due to the smaller size and high mobility. Coating with PPy on textile surface is possible by in-situ chemical polymerization in aqueous solution in presence of oxidant and dopant. In presence of $FeCl_3$ as oxidant, the in-situ polymerization reaction is initiated by the oxidation of pyrrol monomers into radical cations (Figure 3), which combine to form bipyrrole. Continuation of the process leads to formation of polypyrrole (PPy), which is insoluble in water. As a result, PPy precipitates from the solution and in the process part of them are deposited on the surface of the textiles by adsorption. This intercalation of anion of Cl^- has ability of cationic exchange which are present in waste water. PPy exhibits good prospects in adsorption applications because of the nitrogen atoms present in the polymer chains as shown in Figure 3. [8]

Therefore, waste water treatment is one of the novel applications of conducting polymers, particularly of PPy. Various researchers have tried in many ways to remove heavy metal ions from contaminated water by adsorption using polypyrrole prepared via the chemical oxidative polymerization of pyrrole in the presence of different substrate and recipe. The ion adsorption efficiency is found to be largely depends on the substrate on which PPy is deposited and conditions of in-situ synthesis of PPy onto the substrate. [10]

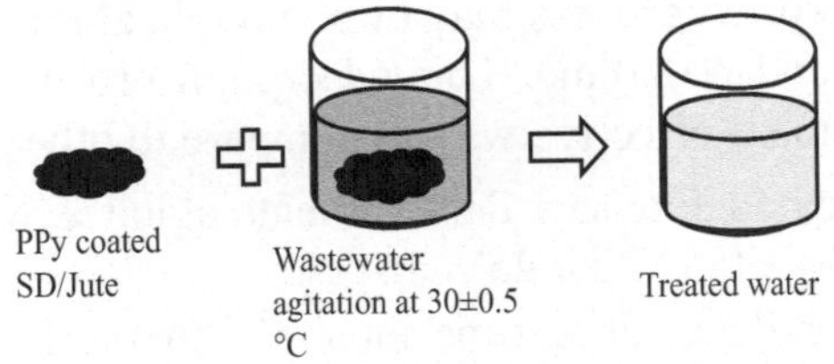

Fig. 2: Chromium Removal Process from Contaminated Water by PPy Coated Bio-adsorbents

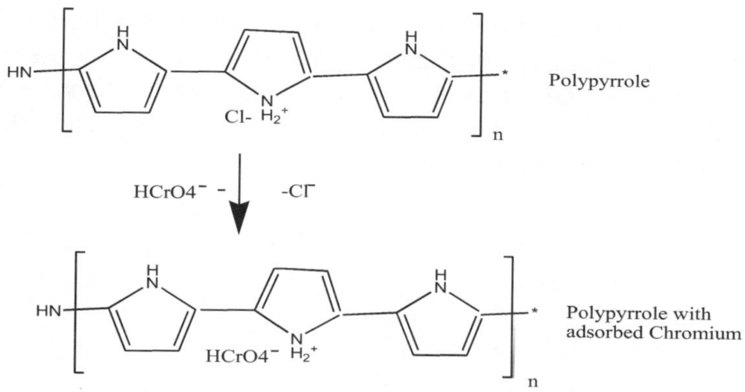

Fig. 3: Scheme of Adsorption of Metal Ion on the Surface of Nitrogen Functional Groups in PPy

3.2 PPy Coated Wood Saw Dust and Jute Fabrics for Chromium Removal from Contaminated Water

Several experiments are conducted on the removal of Cr(VI) ions from water by PPy/SD and PPy/jute composites. It is observed that these composites have considerable potential for the removal of Cr(VI) ions from water using a batch method. In the batch method, the maximum removal efficiency is reported to be at least 99% and 97% by the dose of 1.0 g PPy/SD and PPy/Jute in 25 mL water containing 10 ppm Cr(VI) within 15 min at pH 5.0 and 25°C temperature, respectively as shown in Table 1. The Cr(VI) removal is proposed to be mainly due to the anion exchange properties of PPy which is formed as PPy^+/Cl^- where Cl^- ions which are intercalated in PPy backbone responsible for the ion exchange mechanism as demonstrated in Figure 4. [10]

Table 1: Cr Removal Efficiency of PPy Coated Saw Dust and Jute

Sample	*Initial Concentration of Cr VI in Water*	*Conc. of Cr VI after Treatment with Sample*	*Cr VI Removal Efficiency*
PPy coated Jute	10 ppm	0.03 ppm	97%
PPy coated Saw Dust	10 ppm	0.01 ppm	99%

3.3 UV Spectroscopy Result of PPy Coated SD

UV spectroscopy study was conducted to understand the chromium removal capacity of the PPy coated bio composites. The results are shown in Figure 4.

In case of untreated water sample absorption peaks are obtained at 300 and 370 nm for the presence of Cr III and Cr IV respectively. After the treatment with PPy coated SD, the height of peak at 300 nm is reduced and peak at 370 nm is vanished. These observations imply that Cr VI has been adsorbed from water by the PPy coated absorbents.

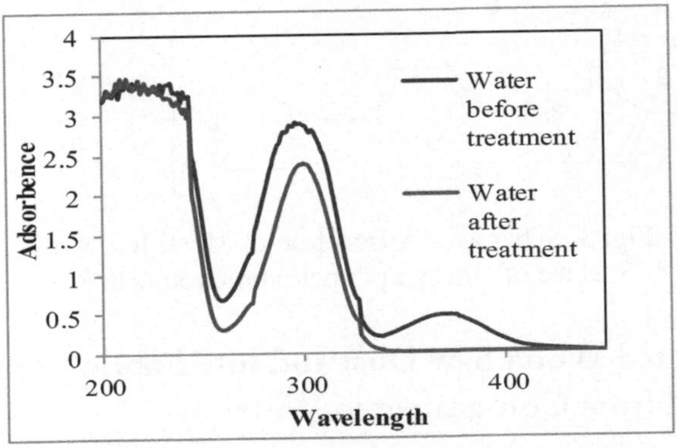

Fig. 4: UV Spectroscopy Study of Untreated and Treated Water

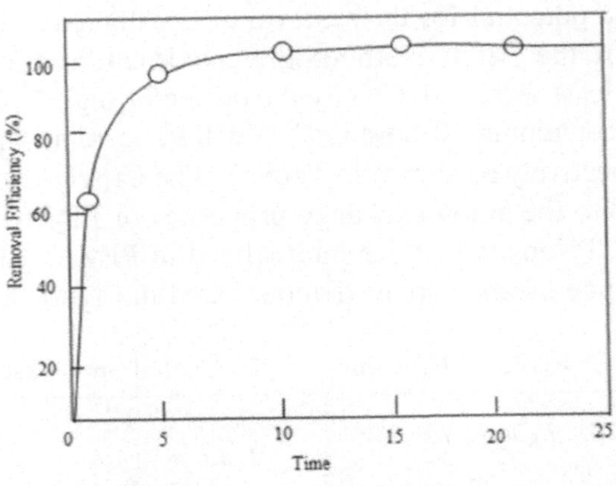

Fig. 5: Adsorption Study of PPy/Jute Bio-Adsorbents for Removal of Cr(VI) from Water

3.4 Adsorption Kinetics

Effect of time for the adsorption of the chromium ions from water had been studied. 1 g PPy/Jute bio-adsorbent was immersed into 100 ml of Cr solution with at pH 5 for different treatment times (0–25 min). The Cr(VI) removal efficiency with time is measured and results are plotted as shown in Figure 6. It can be seen that after 5 min the removal efficiency reaches maximum

(97%) which then sustains even after 25 min. Therefore, minimum 5 minutes of time required for removal of chromium from wastewater.

3.5 Regeneration Process of the Adsorbents

The capacity of adsorption of PPy/Jute sorbents for Cr (VI) reduced drastically after one cycle of treatment. The concept behind regeneration of these bio-adsorbents are to regains their capacity of ion absorbency so that they can be reused and their life cycle can be extended. Researchers proposed three chemical treatment process of regeneration till date based on the desorption mechanism of adsorbed metal ions. These three regeneration operations are treatment PPy/Jute composites using 0.1 M NaOH solution. The alkali treatment process can regenerate about 87% capability of the adsorbent as shown in Table 2.

Table 2: Regeneration Efficiency of PPy/Jute Bio-Adsorbent

Sample	*Concentration of Cr (VI) in Untreated water*	*Concentration of Cr (VI) in PPy/Jute Treated Water after First cycle*	*Concentration of Cr (VI) in PPy/Jute Treated Water after regeneration*	*Regeneration Efficiency (%)*
Regenerated by NaOH treatment	10 ppm	0.03 ppm	0.13 mg/l	87%

4. CONCLUSIONS

The PPy coated jute/SD sorbents are very simple to prepare and can be used as effective removal of heavy metals from wastewater. PPy coated jute and saw dust shows 97% and 99% Cr removal efficiency respectively from aqueous solution measured by MPAES. The adsorption capacity of sorbents can be regenerated by alkali simple treatment. These bio-sorbents can be used to treat the wastewater for removal of harmful metals from water and make water suitable for use in irrigation and municipality. This wastewater treatment will be useful for other industries like paper, textile etc. to reduce contamination in water. Low energy consumption, less time consuming with low operational cost.

REFERENCES

[1] Maity, S., Dubey, A. and Chakraborty, S. (2019). "A Review on Polypyrrole Coated Bio-Composites for the Removal of Heavy Metal Traces from

Wastewater". *Journal of Industrial Textiles*, Published online ahead of print, DOI: 10.1177/1528083719871272.

[2] Baysal, A., Ozbek, N. and Akman, S. (2013). "Determination of trace metals in waste water and their removal processes". In F.S. García & L. Carlos (Eds.), Waste Water - Treatment Technologies and Recent Analytical Developments. IntechOpen. https://doi.org/10.5772/52025

[3] Ghorbani, M., Esfandian, H., Taghipour, N. and Katal, R. (2010). "Application of polyaniline and polypyrrole composites for paper mill wastewater treatment". *Desalination*, 263(1), 279–284. DOI: 10.1016/j.desal.2010.06.072

[4] Katal, R., Ghiass, M. and Esfandian, H. (2011). "Application of nanometer size of polypyrrole as a suitable adsorbent for removal of Cr (VI)". *Journal of Vinyl and Additive Technology*, 17(3), 222–230. https://doi.org/10.1002/vnl.20287

[5] Kumar, P.A., Chakraborty, S. and Ray, M. (2008). "Removal and recovery of chromium from wastewater using short chain polyaniline synthesized on jute fiber". *Chemical Engineering Journal*, 141(1–3), 130–40.

[6] Maity, S. (2019). "Polypyrrole coated Functional Adsorbents for Removal of Heavy metals from Wastewater". International Conference on Advances in Polymeric Materials & Human Healthcare (APA-Stermi), *Asian Polymer Association (APA) in collaboration with Society of Tissue Engineering & Regenerative Medicine India (STERMI),* Goa, India.

[7] Maity, S., Singha, K. and Pandit, P. (2021). "Textile wastewater management. In Green Chemistry for Sustainable Textiles". *Woodhead Publishing*, 417–427.

[8] Maity, S., Singha, K., Pandit, P. and Maiti, S. (2021). "Conjugated polymer-coated novel bioadsorbents for wastewater treatment". Sustainable Technologies for Textile Wastewater Treatments, *Woodhead Publishing*, 157–185.

[9] Singha, K., Maity, S., Pandit, P., Maiti, S. and Lakshmanan, S.O. (2021). "Nanotechnologies for wastewater treatment". Sustainable Technologies for Textile Wastewater Treatments, *Woodhead Publishing*, 1–12.

[10] Pandit, P., Singha, K., Maity, S., Maiti, S. and Kane, P. (2021). "Treatment of textile wastewater by agricultural waste biomasses". Sustainable Technologies for Textile Wastewater Treatments, *Woodhead Publishing,* 137–156.

[11] Maiti, S., Kane, P., Pandit, P., Singha, K. and Maity, S. (2021). Zero liquid discharge wastewater treatment technologies". Sustainable Technologies for Textile Wastewater Treatments. In S.S. Muthu (Ed.), *Woodhead Publishing*, 209–234.

Effect of Pre-Treatment and Dyeing Parameters on Dyeing of Meta-Aramid Yarn

Komal Kukreja and Prasanat K. Panda*
The Bombay Textile Research Association, Ghatkopar West, Mumbai, India
E-mail: nanolab@btraindia.com

ABSTRACT: *The meta-aramid fiber is well accepted raw material for the high-performance applications. Uses of products include apparel fabrics to protect against flash fire and electric arc exposure firefighter garments, filtration applications, insulation in flame-resistant (FR) thermal protective apparel, rubber reinforcement, and in transportation textiles such as aircraft carpeting. For the industrial applications there is no need of dyeing the fiber but for the garment, there is need to dye the material. This is difficult to dye the fiber due to its high crystalline and compact structure. In this study aramid yarn is pre-treated with solvent Dimethyl formamide to facilitate the dyeing process and then dyeing is done with cationic dye. Pre-treatment and dyeing were done at different temperatures and time to see how these parameters affect the process. The fastness properties of dyed fibers found better at high dyeing temperature and time of 140°C and 150 minutes respectively followed by pre-treatment temperature of 90°C and time 60 min.*

1. INTRODUCTION

Aramid fibers are aromatic polyamides. They were developed by DuPont in early 1960s. Commercially meta aramid fibers are known as Nomex. These fibers are being used for making high electrical and thermal protective apparels because of high strength, high modulus and heat resistance property. These fibers can be dyed in different colours through dope dyed process. When small batches of dyeing in different shades are required, exhaust dyeing need to be done. But exhaust dyeing of these fibers is difficult due to its thermostable crystalline structures having high degree of molecular orientation in polymer chain. They also opted strong hydrogen bonding between amide groups in adjacent chains. Various attempts possess made to dye aramid fibers. One feasible method is treatment of fibers with polar solvents such as dimethyl formamide, dimethyl acetamide, and dimethyl sulfoxide which helps in structure opening and easy penetration of dyes. [1] Islam *et al.* [2] used

N-methyl formanilide as swelling agent for cationic dyeing of meta-aramid fibers. It has demonstrated increase in dyeability. Han *et al.* [3] did acid dyeing of meta-aramid yarn. They pre-treated it with PEO_{45}-MeDMA diblock copolymer derived from [2-(methacryloyloxy) ethyl] trimethylammonium chloride. The pre-treatment created dyeing sites which have the ability bond with anionic dyes. Para-aramid fibers can be dyed with disperse dye as shown by Vassiliadis *et al.*, [4] however exhaustion was lower than 79%. Oiwa *et al.* [5] have discussed a method of dyeing of aramid fibers with sulphur or vat dye after a treatment with a polar solvent. Azam *et al.* [6] reported shown dyeing behaviour of cationic dyes at different parameters. Morris *et al.* [7] have shown the effect of pretreatment of the para-aramid fabric with soyabean oil and nonthermal plasma on cationic dyeing. In the present study, aramid fibers are pre-treated with dimethyl formamide and dyed with cationic dye. Systematic study on the effect of pre-treatment and dyeing parameter (temperatures and time) has been done. Effect of change in such parameters in colour value when exposed to light and washing is also studied.

2. MATERIALS AND METHODS

2.1 Material

100% Nomex yarn of count 3/30 was procured from Arvind Limited. Dimethyl formamide (99.8% purity), sodium carbonate and sodium dithionite was procured from Merck Chemicals. Coralene red cationic dye was procured from ColourTex.

2.2 Pretreatment of m-Aramid Yarn

M-aramid yarns were pre-treated with DMF with MLR 1:15 in a water bath. To study the effect of pretreatment temperatures, temperature was varied as 45°C, 65°C and 90°C. To study the effect of pretreatment time, it was done at 15, 60 and 120 minute. Samples were then hot washed for 15 minutes followed by normal wash and open air drying.

2.3 Dyeing of M-aramid Yarn

Dyeing of pretreated aramid yarn was done with 3% shade, MLR 1:25, pH 3–4 at a definite temperature for a definite time. To study the effect of dyeing temperatures, temperature was varied as 100°C, 120°C and 140°C and to study the effect of dyeing time, it was varied as 45, 90 and 150 minute.

Dyed samples were then after treated with 1 gpl solution of sodium carbonate and sodium dithionite at temperature 100°C for 15 minutes followed by normal wash and open air drying.

2.4 Testing of Prepared Samples

Samples were exposed to Xenon arc light of energy 500 W/m^2 and change in colour value after 1, 2, 5 and 8 hours of light exposure is evaluated. For analysis of change in colour value after washing, Soap solution is prepared with standard soap of 5 gpl and sodium carbonate 2 gpl. Samples were then subjected to washing with MLR 1:50 at temperature 95°C for 4 hours in laundometer followed by normal wash and open air drying.

2.5 Colour Value Measurement

Prepared samples were then analysed by Macbeth color-Eye 7000 A spectrophotometer to assess their colour values.

3. RESULTS AND DISCUSSIONS

3.1 Effect of Change in Parameters on Colour Value

Effect of different parameters on colour value is represented is Figure 1. The results show that with the increase in pretreatment temperature, pretreatment time, dyeing temperature and dyeing time colour value of the sample increases.

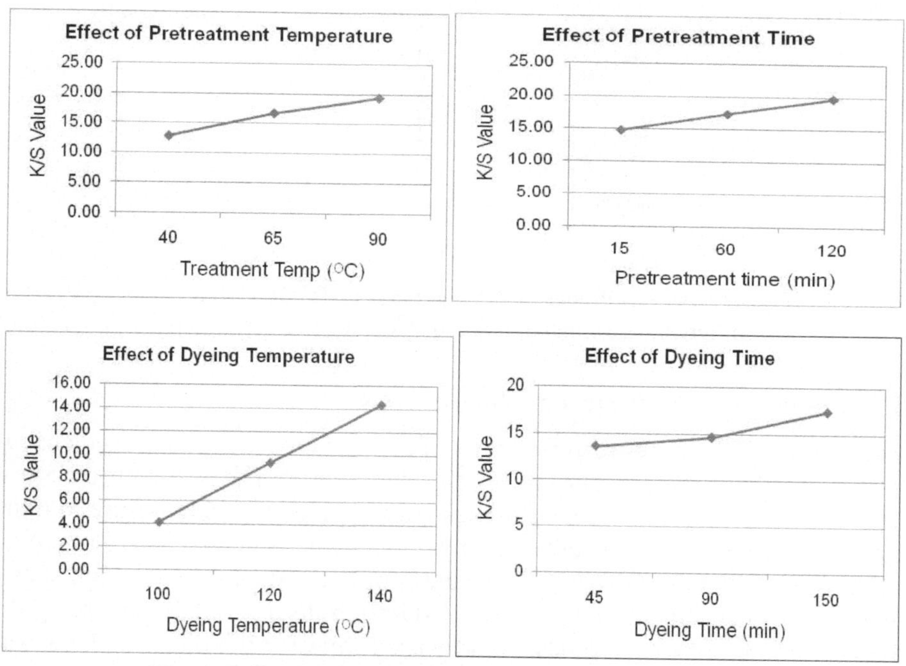

Fig. 1: Effect of Change in Parameters on Colour Value

However, percentage increase in colour value in case of increase in dyeing temperature is higher compared to the dyeing time. This shows that dyeing temperature plays important role in dyeing of aramid fibers. It the effect of pretreatment temperature is more predominated than pre treatment time.

3.2 Effect of Change in Parameters on Colour Value with Light Exposure

3.2.1 *Effect of Pretreatment Temperature*

Effect of change in pretreatment temperature on colour value and percentage change in colour value with exposure to light is shown Figure 2(a) and (b). Colour value of all sample deteriorated with increased exposure time. With the increase in pretreatment temperature, percentage change in colour value after light exposure decreases, which shows that higher pre-treatment temperature helps to improve the fastness to light.

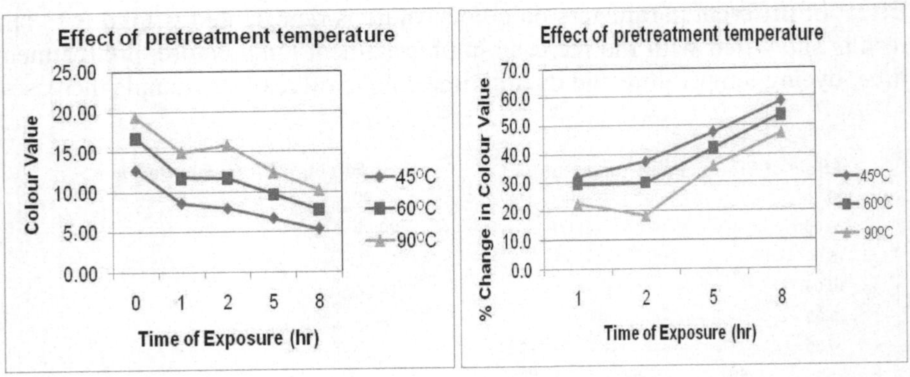

(a) (b)

Fig. 2: Effect of Pretreatment Temperature (a) on Colour Value, (b) in Percentage Change in Colour Value after Light Exposure

3.2.2 *Effect of Change in Pre-treatment Time*

Effect of change in pre-treatment time in colour value and percentage change in colour value with exposure to light is shown in Figure 3(a) and (b). Show decrease in colour value of the samples with the increase in exposure to light. However, with 60 minutes of pretreatment time, percentage change in colour value is less at 15 minutes and 120 minutes of pretreatment time which shows that pretreatment of fibers for a longer duration does not helps in fixing the dye. From the observed values, standardised time for treatment of fiber was kept for 60 minutes.

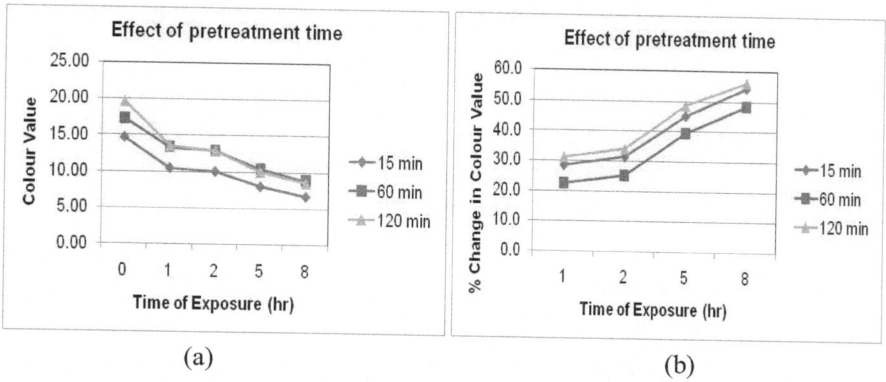

Fig. 3: Effect of Pretreatment Time (a) on Colour Value, (b) in Percentage Change in Colour Value after Light Exposure

3.2.3 *Effect of Change in Dyeing Temperature*

Effect of change in dyeing temperature on colour value and percentage change in colour value with exposure to light is shown in Figure 4 (a) and (b). With the increase in exposure to light, change in colour value and percentage of colour value of all samples shows similar decreasing trend.

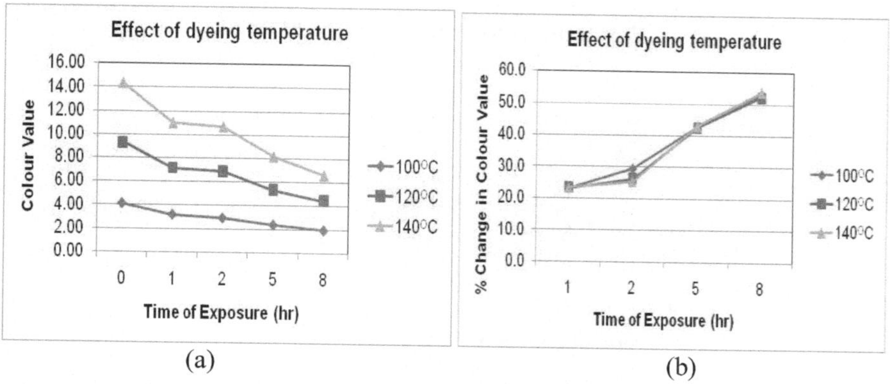

Fig. 4: Effect of Dyeing Temperature (a) on Colour Value, (b) in Percentage Change in Colour Value after Light Exposure

3.2.4 *Effect of Change in Dyeing Time*

Effect of change in dyeing time on colour value and percentage change of colour value with exposure to light is shown in Figure 5(a) and (b). With the increase in exposure to light colour value of all samples dyed at different temperature decreases. There was no significant improvement light fastness was observed with increase in dyeing time over 90 minutes.

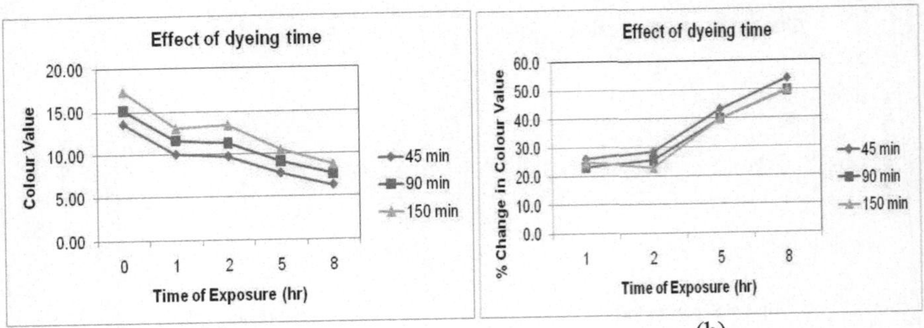

(a) (b)

Fig. 5: Effect of Dyeing Time (a) on Colour Value, (b) in Percentage Change in Colour Value after Light Exposure

3.3 Effect of Parameters on Colour Washing Fastness

3.3.1 *Effect of Pretreatment Temperature*

Effect of change in pretreatment temperature in washing fastness is shown Figure 6. At higher temperatures of 65°C and 90°C, percentage change in colour value is similar.

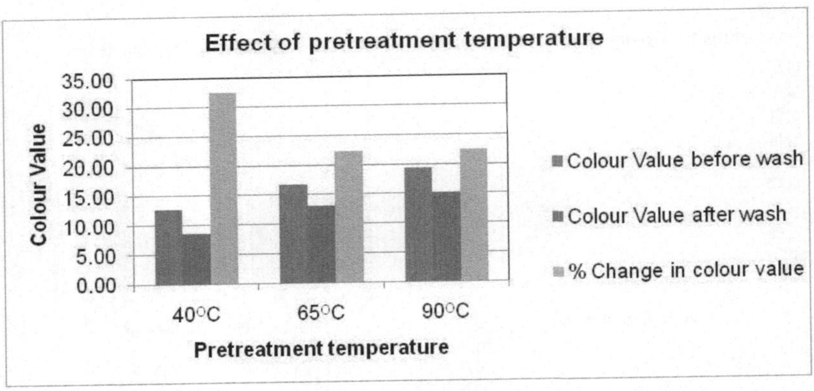

Fig. 6: Effect of Change in Pre Treatment Temperature in Colour Value after Washing

3.3.2 *Effect of Pretreatment Time*

Effect of change in pretreatment time in colour value of samples after washing is shown Figure 7. At 60 minutes of pretreatment, percentage reduction in colour value is less as compared to 15 minutes and 120 minutes of pretreatment. It is evident that 60 minute pre treatment time is optimum. This result is an agreement to change in colour value to light exposure.

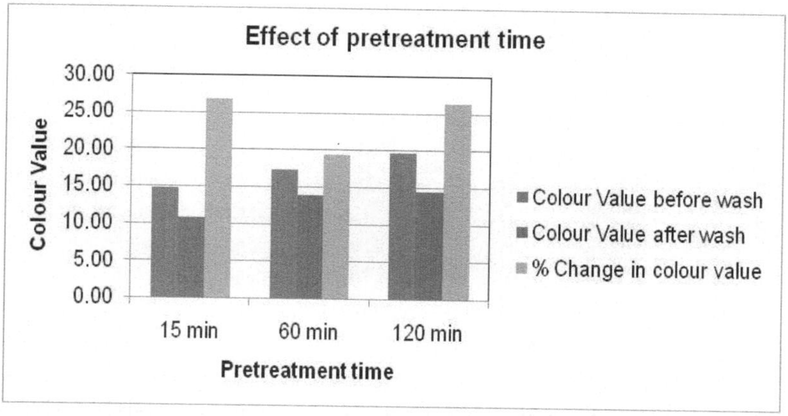

Fig. 7: Effect of Change in Pre Treatment Time in Colour Value after Washing

3.3.3 *Effect of Dyeing Temperature*

Effect of change in dyeing temperature time in colour value of samples after washing is shown in Figure 8. At temperature 140°C percentage reduction in colour after washing was observed compared to low temperature values.

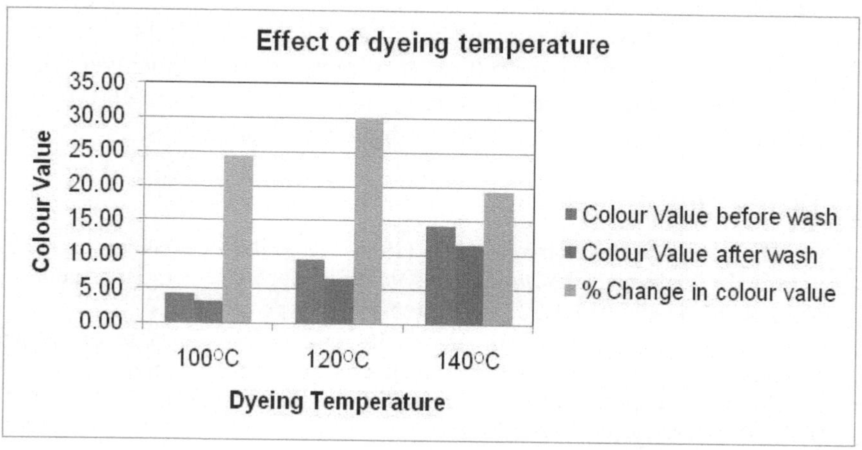

Fig. 8: Effect of Change in Dyeing Temperature on Colour Value after Washing

3.3.4 *Effect of Dyeing Time*

Figure 9 is representing the effect of dyeing time on washing fastness. At 150 minutes of dyeing time, percentage reduction in colour was found less after washing.

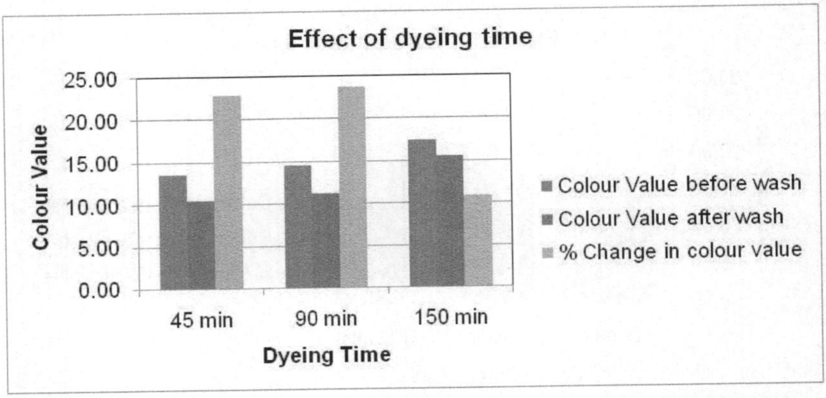

Fig. 9: Effect of Change in Dyeing Time in Colour Value after Washing

4. CONCLUSION

In this study, effect of pretreatment temperature, pretreatment time, dyeing temperature and dyeing time on dyeing of aramid fibers was studied. Effect of dyeing temperature is more prominent than other parameters. Higher pretreatment temperature with standard pretreatment time and higher dyeing time leads to darker shade. In this study standard parameters were pretreatment at 90°C temperature for 60 minutes with dyeing at 140°C for 2.5 hours. These results were supported by the percentage colour change after light exposure and washing. Higher dyeing temperature & prolonged time positively affects washing fastness. At the pretreatment temperature 90°C and time 60 min there is comparatively less fading of samples to light and wash fastness.

REFERENCES

[1] Moore, R.A.F. and Weigmann, H.D. (1986). "Dyeability of Nomex® Aramid Yarn". *Textile Research Journal*, 56, 94–99. https://citeseerx.ist.psu.edu/viewdoc/download?doi=10.1.1.823.2753&rep=rep1&type=pdf

[2] Islam, M.T., Aimone, F., Ferri, A. and Rovero, G. (2015). "Use of N-methyl formanilide as swelling agent for meta-aramid fibers dyeing: Kinetics and equilibrium adsorption of Basic Blue 41". *Dyes and Pigments*, 113(2), 554–561. DOI: 10.1016/j.dyepig.2014.08.029

[3] Han, S.Y. and Jaung, J.Y. (2009). "Acid Dyeing Properties of Meta-aramid Fiber Pretreated with PEO45-MeDMA Derived from [2-(Methacryloyloxy)ethyl] Trimethylammonium Chloride". *Fibers and Polymers*, 10, 461–465. doi: 10.1007/s12221-009-0461-6.

[4] Vassiliadis, A.A., Roulia, M. and Boussias, C.M. (2006). "Disperse dyeing systems for p-aramid fibers". *37th International Symposium on novelties in Textiles, Ljubljana, Slovenia.*

[5] Oiwa, N., Imai, K., Sayama, S., Ito, T. and Yasui A. (2014). "Method for dyeing aramid fibers and dyed aramid fibers". *United States Patent Application Publication,* US 2014/0020190 A1, 1–21.

[6] Azam, F., Iqbal, K., Safdar, F., Hussain, T., Ashraf, M. and Rehman, A. (2019). "An investigation into the effect of different parameters on the dyeing of high-performance m-aramid fiber and its optimization". *Clothing and Textiles Research Journal,* 38(2), 0887302X1987087. DOI: 10.1177/0887302X19870872

[7] Morris, M., Ye, X.P. and Doona, C.J. (2021). "Dyeing para-aramid textiles pretreated with soybean oil and nonthermal plasma using cationic dye". *Polymers,* 13(9), 1492. https://doi.org/10.3390/polym13091492

Green Technologies for Textile Processing

Munjal Parikh

Man-Made Textiles Research Association, Ring Road, Surat, India

E-mail: director@mantrasurat.org

ABSTRACT: *Large number of hazardous chemicals which includes dyestuffs, printing auxiliaries, finishing agents, and other auxiliaries are being used in textile wet processing. After processing, all these chemicals are discharged as effluents which consequently create pollution, which ultimately environment and pose serious threats to human health. Although effluent treatment techniques are adopted to minimize the discharge of hazardous substances into the environment it is imperative that pollution generation at source is reduced. Green approach to textile processing means use of green fibres, chemicals, solvents and to optimize the processes, adopting reuse and recycling of process waters. It also includes use of green fuels and it is combination of all above mentioned approaches. This paper reviews various technologies and approaches used in processing of textile fabrics to achieve sustainable production. It also discusses latest trends adopted by industry.*

Keyword: Textile Wet Processing, Unfixed Dyes, Pollution, Green Fibres, Sustainable Production.

1. INTRODUCTION

Textile processing industry is known to produce pollution as it has many processing steps that uses large amount of water, dyes and chemicals. When the process is over the spent water containing unfixed dyes, chemicals are drained which creates waste water or effluent. Green chemistry, also known as sustainable chemistry," materials produced in this process nither use or creates hazardous compounds" according to definition. [1] To achieve this we have to adopt textile processing sequence and methods which result in no or at least minimum waste production. As a part of greener strategy, it's better to prevent waste production rather than to go for treatment strategies to dispose it off.

2. DYES AND CHEMICALS THAT CONTRIBUTES TO POLLUTION

The conventional textile wet processing has basic order as follows.

1. Pre-treatment
2. Dyeing
3. Printing
4. Finishing

Pre-treatment includes desizing, scouring, bleaching and whitening. The removal of size from warp yarn is known as desizing. [2] If starch is used then it is removed by hydrolysis by oxidation or by enzymes. Due to these broken fragments BOD of waste water is increases. Also, PVA based starch is removed by hot water aided by wetting agent. Here both will add to BOD/COD of effluent. In scouring generally alkali in the form of caustic soda is used with detergent, wetting agents, emulsifiers and solvents present in stain remover. Scouring removes dirt, wax and oils, etc. The scouring bath will contain detergents, wetting agents, etc plus the impurities removed that all will contribute to COD and BOD of effluent. The NaOH will increase TDS of the effluent.

Table 1: Process and Pollutants (Polyester Fabrics)

Process	*Pollutants in Drain*
Scouring & bleaching	Detergent, alkali, solvents, wetting agents, sulphur compounds, dirt, oil, grease
Washing (two times)	Residue of above
Weight reduction	Residual alkali, PTA, Ethylene glycol
Washing (two times)	Residue of above
Dyeing	Unfixed dye, acid, leveling agent, wetting agent
Washing	Residue of above
Reduction clear	Byproducts of hydro, alkali, dye fragments
Printing	Gum, unfixed dye, printing chemicals
Washing after steaming	Byproducts of hydro, alkali, dye fragments, gum, printing chemicals

The bleaching is the process which destroys colour from the fabric and makes it white. In cotton generally oxidising agents are used while for synthetic fiber fabrics sodium hydrosulphite is used. In case of cotton the excess of bleaching agents adds to the pollution whereas in hydros its decomposition products contribute to COD of waste water. In case of cotton mercerisation is done with strong alkali which contributes to high pH and increases TDS.

In dyeing the dyes are generally applied from aqueous bath and at the end of dyeing most of the dyes get fixed inside the fiber the extent of which depends on type of dye. Dyeing process uses variety of chemicals like levelling agents, pH buffers, salts, etc. After dyeing unfixed dyes and all auxiliary chemicals remain in spent bath. These leads to increase BOD, COD and TDS of the effluent.

During printing to promote fixation of the dye, thickener is added with auxiliary chemicals. When printed fabric after ageing is washed entire amount of thickener along with auxiliary chemicals come out into washing liquor it also contains unfixed dye. Thickeners add to the BOD of waste water. The rest of chemicals, unfixed dye as well as detergent used for washing add to COD of waste water. In case of polyester printing with disperse dye reduction clearing is given with the help of sodium hydrosulphite and caustic which adds to TDS and toxic sulphur compounds.

In finishing steps the left over finishing auxiliaries are directly drained off which may contain toxic compounds like formaldehyde based resins, anti-microbial agents and so on. All these pollutants

- Decrease Dissolve Oxygen content (DO) and increase pH,
- Total Suspended Solids (TSS),
- Total Dissolved Solids (TDS),
- Chemical Oxygen Demand (COD), and
- Biological Oxygen Demand (BOD).

The other area of concern is the usage of water and energy. Energy is used for imparting heat to the machines which is generated in boilers. It requires non renewable sources of fuels like coal, lignite, natural gas etc. In order to apply green technologies it is required to minimize use of fossil fuels and use alternate source of energy. Textile processing also uses huge quantity of water at each stage. This ultimately uses more energy and increases quantum of waste water generated. Therefore savings on water consumption is essential for green approach.

3. GREEN APPROACH IN PROCESSING

The term "sustainability" is used to refer to the degree to which processes or products help to protect the environment for future generations. Sustainable development satisfies does not affect or impact to resources for future generations. [2]

We have already discussed the points of concern that makes textile processing un-sustainable. In order to make textile processing sustainable we will have to adopt following strategies.

1. Selection of chemicals and dyes
2. Use of renewable energy
3. Minimize water and energy consumption
4. Design process to create less pollution
5. Adopt pollution prevention approach

Selection of chemicals and auxiliary chemical is important. Chemicals should be less hazardous to environment. It should be biodegradable and non toxic. e.g. Use of formic acid is preferable over acetic acid as shown in Table 2. Formic acid exerts less pollution load and is preferred.

Table 2: Equivalent BOD for Acetic Acid and Formic Acid

Acid	*BOD Equivalent kg/kg*
Acetic	0.64
Formic	0.12

Another example is selection of detergent such as Nonyl phenol ethoxylate, which is cleaning agent also. However it has poor biodegradability. Further when degraded it release nonyl phenol which is toxic to fish. So we should replace it with fatty alcohol ethoxylate which are relatively safer for the environment. Chlorinated solvents should be avoided for stain removing products and it should be replaced with alcohol based solvents. Before using any chemical or auxiliary its MSDS should be studied and compared. Ideal chemical should be biodegradable and should have relatively low COD. Dye selection needs to be focussed on high fixation efficiency and its decomposition products should be safe for receiving waters. There has been continuous research going on to improve dyeing property of dyes. Use of these can reduce energy consumption. eg. 50% savings in energy and water consumption have been observed with the introduction of new bi- and multifunctional reactive dyes. Dyeing with multifunctional dyes also reduced the addition of additive salts into the dye bath by at least 20%. Now pre-reduced forms of sulphur and vat dyes are available which has simplified the dyeing process particularly for denim production.

Water saving will lead to saving in energy consumption and lower the quantity of effluent generation. Therefore water conservation is first step to achieve sustainability and important part of green technology. Before attempting these one should opt for water audit that gives idea of water consumption pattern. Reuse of process water is highly recommended where ever possible. There are possibilities where we can adopt this in pre-treatment and dyeing steps.

The process sequence can be designed in such a way so that process becomes sustainable. In polyester fabric pre treatment scouring and bleaching are combined which reduces amount of water and energy used. Sometimes it is possible to combine even desizing, scouring and bleaching in one step to save huge amount of water and energy. [4] Use of pad batch method of reactive dyeing with high reactive dye in cold can be a better option than batch wise dyeing in jigger.

"The method which limit or eliminate the formation of pollutants due to usage of material or process "is known as pollution prevention (also known as source reduction). [3] Reduction of pollution at source can be achieved through reduce chemicals, Water, energy as already discussed. It can also be achieved through reuse/recycle which minimizes introduction of fresh raw materials. Recycle waste process water into another useful application is also possible.

4. RECENT TRENDS IN GREEN TECHNOLOGIES

The other discussed green dyeing process is supercritical carbon dioxide (SCO_2) dyeing, where instead of water supercritical carbon dioxide is used. It can be used to dye polyester and polyamide using disperse dyes. [5] Supercritical Carbon dioxide dissolve disperse dyes at high temperature and pressure due to their high diffusion rate and low viscosities. Dye diffusion takes place efficiently and the end of dyeing unutilized dye powder is recovered. So there is no need for drying and no effluent is generated. However this technology requires a vessel to create very high pressure to bring CO_2 gas under supercritical conditions.

Green technology also demands use of green fiber in manufacturing of textiles. Silk, organic wool, organic cotton and jute are typical green fibers. On manmade fibers there are poly lactic acid and lyocell fibers. Lyocell is type of a regenerated fiber like viscose but produced with eco friendly route. PLA is obtained from corn oil is a biodegradable fiber. Poly terpthalate PTT fiber is now available as partly renewable resourced fiber where one of the raw material is obtained from corn oil.

Much importance is also given to application of Plasma technology as green technology in textile processing. [7] Plasma generates highly charged atmosphere consisting of electrons, negative and positive ions, radicals, and atoms. Fabric surface gets modified due to the following points:

- physiochemical changes depending on the nature of gaseous matter,
- type of fabric,

- current,
- pressure, and
- temperature.

Dyeability, wettability, pilling, and water resistance characteristics of textiles changed as the surface characteristics changes. [6]

Under specific condition plasma generates desired functionalities on textile surface. It also remove impurities from surface, can produce surface etching etc. Although much is discussed about plasma application in textiles extensive research work is going on worldwide and very soon this will become commercially successful green technology.

5. CONCLUSION

With pollution control authorities becoming more stringent textile industry will have to move from conventional approach to green technology. Use of environmentally safe dyes and chemicals and optimizing and redesigning of process, reducing water and energy is only possible if due attention is given to green technologies.

REFERENCES

[1] Chowdhury, A.K.R. (2013). "Green chemistry and the textile industry". *Textile Progress*, 45(1), 3–143. https://doi.org/10.1080/00405167.2013.807601

[2] Hussain, T. and Wahab, A. (2018). "A critical review of the current water conservation practices in textile wet processing". *Journal of Cleaner Production*, 198, 806–819.

[3] Cannon, A.S., Finster, D., Raynie, D. and Warner, J.C. (2017). "Models for integrating toxicology concepts into chemistry courses and programs". *Green Chemistry Letters and Reviews*, 10(4) 436–443. DOI: 10.1080/17518253.2017.1391880

[4] Franca, C.C.V. and Ueno, H.M. (2020). "Green cosmetics: perspectives and challenges in the context of green chemistry". *Desenvolvimento e Meio Ambiente*. 53, 133–150. DOI: http://dx.doi.org/10.5380/dma.v53i0.62322

[5] Zhao, W. (2018). "Make the chemical industry clean with green chemistry: an interview with Buxing Han". *National Science Review*, 5(6), 953–956. https://doi.org/10.1093/nsr/nwy045

[6] Nosonovsky, M. and Bhushan, B. (2010). "Green tribology: principles, research areas and challenges". *Philosophical transactions. Series A, Mathematical, physical, and engineering sciences,* 77–94. https://doi.org/10.1098/rsta.2010.0200

[7] Rout, P., Puspendu, B., Ramakrishnan, A., Rao, S. and Tyagi, R. (2016). "Sustainable Hazardous Waste Management/Treatment: Framework and Adjustments to Meet Grand Challenges". *Sustainable Solid Waste Management*. DOI: 10.1061/9780784414101.ch12

Application of Indian Madder Dye on Woolen Fabric Sample and Assessment of Fastness Properties

Sushanta Naik
Shri Vaishnav Institute of Textle Tcehnology, SVVV, Indore, MP, India
E-mail: sushantanaik@svvv.edu.in

ABSTRACT: *The dyeing of woolen fabric sample was carried out by aqueous extract obtained from ground Indian Madder (Manjistha) root with and without using mordant. Pre mordanting method was used for the woolen fabric samples dyed using mordants. The mordants used for this purpose were potash alum and iron sulfate. Dyeing was carried out in mild acidic condition using acetic acid at 90^{0}C temperature. The effect of mordant on color strength, light fastness, washing fastness and rubbing fastness properties has been investigated. Bright red, dark red, and black colour shades were obtained by using different mordants. Improved fastness properties are also observed as a result of pre mordanting of woolen fabric samples with potash alum and ferrous sulfate.*

Keywords: Pre-mordanting, Color Fastness, Natural Dye, Potash Alum, Indian Madder.

1. INTRODUCTION

The use of Indian madder as a dye is an age-old process in India. Since the last quarter of eighteenth century, India was the biggest exporter of dyed fabrics in the world. The dyed clothes of India were sold at premium prices in Europe. After arrival of synthetic dyes in the market, the natural dyes like Indian madder became obsolete. Generally the fastness properties of natural dyes like washing and light fastness are poor. So, synthetic dyes became more popular for their good colour fastness effect. However their manufacturing and application impart hazardous effect to the environment. Eco-friendly products and green chemistry are gaining popularity in present era and so it has become necessary for textile chemists to discover eco friendly alternatives of synthetic dyes and chemicals. So many trails have been undertaken through research for the use and application of eco- Friendly dyes around the world. [1,2]

Indian madder (Rubia cordifolia) comes from the family of Rubiaceae. This is a herbaceous plant with a long root. The most important part of Indian madder plant is its root. This is the source of natural dye. The main coloring compound present in Indian madder are purpurin, munjistin, xanthopurpurin and and pseudopurpurin. For cotton, silk and wool madder is one of the oldest and predominant natural dyes. Mordants are used to create affinity between textile and natural dye and to improve fastness of natural dyed textile materials. Mordants generally contain polyvalent metal ions. They form coordinate bond with fiber and coloring component of natural dye. [1]

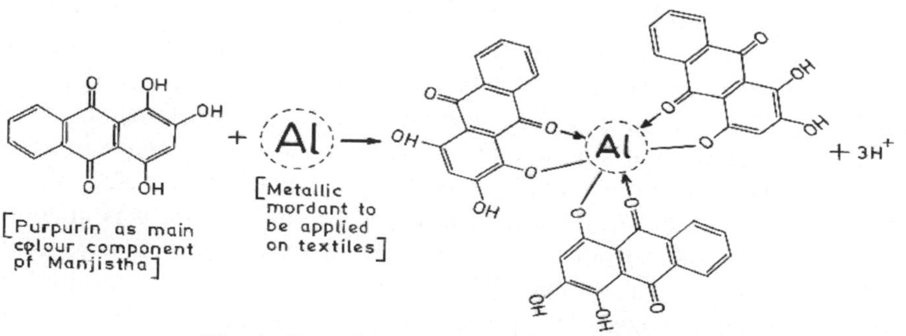

Fig. 1: Co-ordinate Bond Formation Reaction between Idian Madder and Mordant [1]

Some of the important examples synthetic mordants are potash alum, potassium dichromate, ferrous sulfate and stannous chloride. They are responsible for improving dye shade and fixation properties.

In this study, woolen fabric samples are dyed with Indian madder dye using potash alum and ferrous sulfate mordant. The dye sample fastness properties were examined on woolen dyed fabric.

2. MATERIALS AND METHODS

2.1 Fabric, Mordant and Dye

100% pure woolen fabric was purchased from Wool Research Association, Mumbai and used for this study. Indian madder, which was commercially available, was used for dyeing. Laboratory grade potash alum, ferrous sulphate, hydrochloric acid and acetic acid were used for this study.

2.2 Mordanting with Potash Alum and Ferrous Sulfate

Potash alum and ferrous sulphate were used in pre mordanting process. In different dye pots, 5% ferrous sulfate (on the weight of fabric) and 10% potash alum (on the weight of fabric) were mixed in water by continuous stirring. The ratio of material and liquor is 1:30. The temperature of the dye-bath is set at

30°C and woolen fabric was added in the dye pot and the temperature of the bath is raised up to 90°C. This mordanting process was carried out for 1 hour. Then the baths were cooled. The mordanted fabric were taken out from dye bath and then rinsed with flowing water.

2.3 Dye Extraction

The extracted compound of coloring matter was used for dyeing purposes. 30% (on the weight of fabric) of ground Indian madder root was taken in a dye pot. Then it was pasted with Turkey Red oil. Then required amount of water was added in the dye pot and continuously stirred. Acetic acid was added to keep the pH 4–5. The ratio of material and liquor is 1:30. Then continuous stirring was carried out for the solution of Indian madder for 12 hours in acidic condition. There after the solution is boiled for 1 hour, cooled and filtered. After extraction of Indian madder root, orange color solution was obtained. The remaining residue from the filter was again mixed with aqueous acidic solution and the same above mentioned process was carried out until all the colors have been extracted. This process was repeated until all the roots were converted to dye solution and until no residues were left. The extracted colorant is a solution of Indian madder dye. Then woolen fabric was dyed between 4–5 PH using this solution. [3,4]

2.4 Dyeing

The dye bath was set using optimum level of water, required dye and acidic acid. The ratio of material and liquor was 1:30. The PH is maintained between 4 to 5 during dyeing process. Dyeing was commenced at 30°C. The temperature of the dye bath is gradually raised up to 90°C and temperature is maintained at this level with continuous stirring. There after dyed material is washed by using 5 ml/l non-ionic detergent. Lastly after washing and through rinsing in water, the samples were dried in shade. [3,4]

2.5 Colour Measurement

Woolen dyed sample is measured by premier color scan spectrophotometer (SS5100A) for getting color strength. Kubelka-Monk equation was used to find out the colour strength (K/S value) of woolen fabric samples dyed with Indian madder dye using different mordants.

The equation is,

$$K/S = (1-R^2)/2R \quad \ldots (1)$$

Where K is coefficient of absorption, R is called reflectance of the sample and S is scattering coefficient.

2.6 Fastness Testing

The fastness properties of dyed woolen fabric samples were analyzed to find out fastness of color for washing, light and rubbing.

2.7 Light Fastness

The XENTEST light fastness testing machine (Xenon are Lamp) was used for the light fastness measurement using standard testing method IS: 2454:1985—RA 2006.

2.8 Washing Fastness

The washing fastness of the dyed woolen fabric samples were measured as per IS/ISO 105 C10 A1:2006 specifications. A layer of dyed fabric (10 cm × 4 cm) was sewn between two pieces of adjacent fabrics (wool and polyester). The specimen was treated with 5 g/l standard soap at 40°C for 30 minutes. The samples were assessed for colour change on washing.

2.9 Rubbing Fastness

Dry and wet rubbing of dyed woolen sample were tested by crock meter as per Indian standard IS 766:1988 (Reaffirmed 2004) based on ISO 105-X12:2001 by mounting the fabric on panel and giving 10 strokes. The samples were assessed for staining on white adjacent cotton fabrics.

3. RESULTS AND DISCUSSIONS

3.1 Colour Measurement

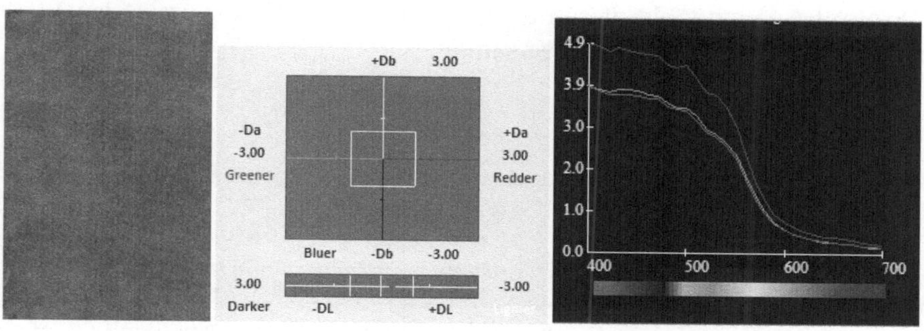

Fig. 2: Unmordanted Wool Fabric Dyed with Indian Madder

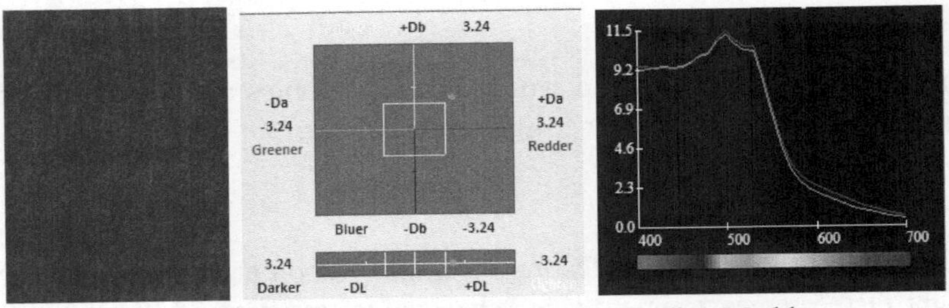

Fig. 3: Wool Fabric Sample Dyed with Indian Madder Dye after Pre-mordanting with Alum

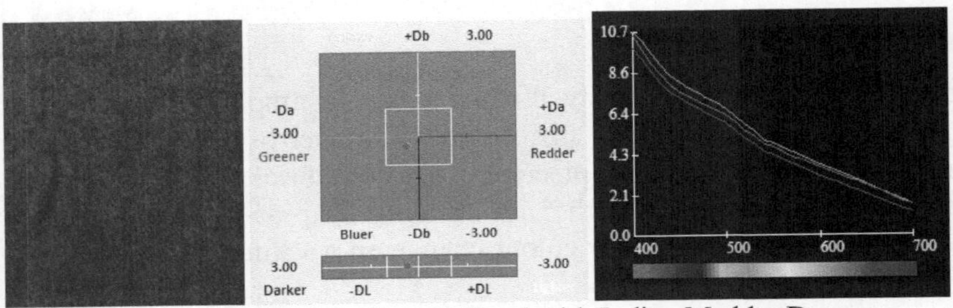

Fig. 4: Wool Fabric Sample Dyed with Indian Madder Dye after Pre-mordanting with Iron Sulfate

Table 1: Color Absorption was Obtained with Indian Madder

Sr. No.	*Shade%*	*Mordant*	*L**	*a**	*b**	*C**	*K/S*
1.	30	Nil	52.03	27.27	24.32	36.54	44.96
2.	30	10% Alum	36.73	25.04	18.64	31.22	116.54
3.	30	5% Iron sulfate	36.43	6.25	12.96	14.39	97.70

From the above table, it can be seen that, mordanted samples have more colour strength than unmordanted samples.

3.2 Fastness Properties

Table 2: Fastness of Unmordanted & Mordanted Woolen Fabric Samples

Sr. No.	*Shade %*	*Mordant*	*Light Fastness*	*Washing Fastness*			*Rubbing Fastness*	
				c.c.	c.s.	c.w.	Dry	Wet
1.	30	Nil	4	2–3	4–5	3–4	5	4–5
2.	30	10% Alum	4	4–5	4–5	5	5	4–5
3.	30	5% Iron sulfate	4	4	4–5	3–4	5	4–5

Note: c.c.- colour change, c.s.- colour staining on cotton, c.w.- colour staining on wool.

3.3 Light Fastness

There was no difference between light fastness of unmordanted and mordant sample of wool. Both have similar resistance to photochemical degradation.

3.4 Washing Fastness

The washing fastness of the dyed woollen samples was increased due to addition of alum and ferrous sulphate. The fixation of dye with fiber is improved by addition of mordant. Very slight staining was observed on adjusant cotton fabric in all set. On adjacent wool fabrics, alum mordanted samples showed no staining, while unmordanted and ferrous sulfate mordanted samples shown slight to very slight staining.

3.5 Rubbing Fastness

In rubbing fastness test of dry & wet samples of unmordanted and mordant fabric, there were no difference in rubbing fastness properties. Both unmordanted and mordanted samples have equal resistance to abrasion caused by crock meter.

4. CONCLUSION

The dye extracted from Indian madder root powder was found to have acceptable dyeing properties producing excellent shades of red and black colour. Significant difference in colour strength of dyed woolen fabric samples as a result of application of different mordants was observed. Alum mordanted samples show higher colour strength as compared to ferrous sulfate mordanted and unmordanted woolen fabric samples. So Indian madder can be used as a sustainable and environment friendly dye to get red and black colour on wool using different mordants.

REFERENCES

[1] Samanta, A.K. and Konar, A. (2011). "Dyeing of Textiles with Natural Dyes". *Natural Dyes*, In Dr. Emriye Akcakoca Kumbasar (Ed.), ISBN: 978-953-307-783-3, InTech, http://www.intechopen.com/books/natural-dyes/dyeing-of-textiles-with-natural-dyes.

[2] Gulrajani, M.L. and Gupta D. (1992). Natural dyes and application to textiles, Department of textile technology, Indian Institute of Technology, New Delhi, India.

[3] Yusuf, M., Shahid, M., Khan, S., Ahmad, Khan M.I., Islam, S-Ul, Faqeer, M. and Khan Ali M. (2013). "Eco-Dyeing of Wool Using Aqueous Extract of the Roots

of Indian Madder (Rubia cordifolia) as Natural Dye". *Journal of Natural Fibers*, 10(1), 14–28.

[4] Feiz, M. and Norouzi, H. (2015). "Dyeing and Adsorption Studies of Madder (Rubia tinctorum) on Wool Fabrics". *Materials Science,* 21(1). DOI:10. 5755/j01.ms.21.1.5150.

Review on Compound Fabrics

Ravikumar Purohit*, U.J. Patil and Ashish Hulle
Department of Textiles, D.K.T.E. Society's Textile and Engineering Institute, Ichalkaranj, Maharashtra, India
E-mail: *Purohitravi49@gmail.com

ABSTRACT: *The fabric which is to be converted into garments is generally made up of woven or knitted structures. Each structure has their own advantages. Woven structure has good strength because of good interlacement while knitted structure has good comfort because of interloping. So, depending on the properties, cost and required performance the fabric is then converted into the required end products. Polyester and cotton each can be converted into finished fabric, and both has their own advantages and disadvantages but mixing of polyester and cotton has given a new direction to all the manufacturer to save some cost and betterment in many properties. Thought individual structure woven and knitted has their own benefits but the study can be done for betterment in the result in many aspects. This betterment can be achieved by structure compounding. So, it is very much important to study the new way of fabric compounding. This review discusses the different structure and their properties that are generally used to manufacture the apparels. Their advantages and disadvantage of each structure, meaning of fabric compounding and how this compounding of the fabric can be done.*

Keywords: Types of Structure, Compound Fabric, Woven and Knitted Structure.

1. INTRODUCTION

Textile fabrics are an essential portion of every human alive on this planet earth. Textile material has good strength, sufficient flexibility, permeability, and drape property. These parameters are important to obtain substantial protection, aesthetics, and comfort all through their use as application in apparels or garments. [1] Tailoring of fabric construction/ materials is always a major ingredient to develop an optimum product. For the right product the appropriate fabric properties are required. Fabric properties are affected by several factors like fibre selection (cotton, polyester, viscous, nylon etc.), type of yarn (spun, fancy, textured, filament etc.), type of structures fabric is be used (woven, weft knitted, warp knitted, nonwoven etc.), and fabric finishing. [2] The behaviour of the fabric and characteristics plays the vital role in the designing of a garment. 100% polyester plain woven with high twist is less smooth while on other hand a warp knitted 100% polyester fabric is used to transfer the moisture from the skin to the at mosphere. It makes fabric everyday wear. Generally, clothes intended for fashion clothing must accomplish many characteristics. These properties are strength, durability, colourfastness, comfort, aesthetics and so forth. They are very important for everyday use. Fabrics for apparels are either woven, weft knitted, or warp knitted. They are composed of most synthetic or natural fibres with different count to provide an optimum performance. Hence fibre properties, structure, finishes and type of different yarn affect the end properties. [3,4]

2. DIFFERENT TYPES OF STRUCTURE USED FOR CLOTHING

Fabric Structure can affect the performance of fabrics. It particularly controls thermo-regulation of human body. Different types of fabric structures are used as per end use in apparel production. The most used structures are woven and knitted but some additionally uses nonwoven, braided and many more. Woven and knitted fabrics perform differently due to their different structural make up which is summarized in the following segments. [1,5,6]

2.1 Woven Structure

Two or more sets of yarns are interlaced to form the fabric. They interlace with each of other at right angles. Woven fabric shows partial elasticity and adaptability to body movement. The weave structure of fabric controls the bulkiness and wrinkle recovery. Woven structure has good stability to stress, and it is also less air permeable, especially with dense fabric. Woven structure has good stability in normal use &care, and it generally shrink lesser than 2%. Woven structure is preferred where we need more breaking strength, tearing strength, dimensional stability and for some special purpose material.

2.2 Knitted Structure

In knitted structure interconnected loops series are present and this is made with eighter one or many sets of yarns. Knitted structure Possess stretch and elasticity, adapts to body movement and it possess open spaces between yarns. Knitted structures demonstrates has good wrinkle recovery and air permeability. They are less stable in handle. They have higher shrinkage if not heat. From application point of view knit is preferred for good drape, moisture permeability and comfort.

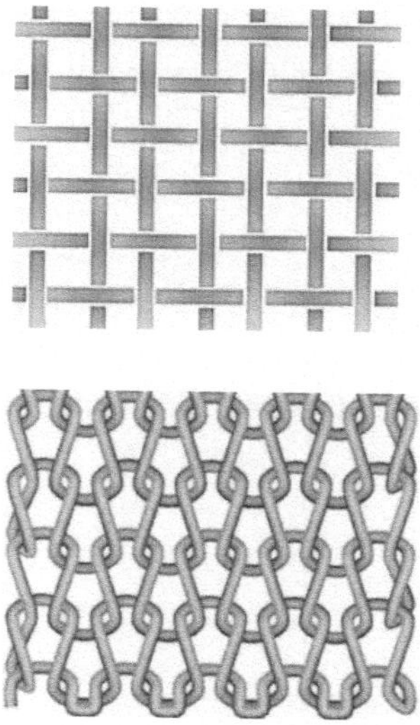

Fig. 1: Woven and Knitted Structure

3. MEANING AND NEED OF COMPOUND FABRIC

Something that consists of two or more things or substances combined is known as compound material. Compound fabric structure means two or more type of structure are combined to form one product. In a single product generally, single structure is used. Depending on the required parameter the structure is chosen, in some cases one product may need more structure. The compound fabric can solve this problem. For the single product the combination may be required like good durability at few places and good crease recovery at some places. So, the compound structure can solve these custom requirements.

3.1 Benefits of Fabric Compounding

1. When two different types of fabric are mixed, the properties of these two fabrics may synergized.
2. Compounding is done to produce a fabric, which is economical by combining the aesthetic comfort properties of one structure and another.
3. Compounding can also help to reduce fabric's weight with all desirable characteristics. [7]

3.2 Ways to Create Compound Fabric

Compounding can be done in various ways if we consider only 2 structure that is woven and weft knitted structure, then also we can have lengthwise, width wise and layer wise compounding. Figure 2 shows different fabric compounding the combination made with woven and weft knitted material.

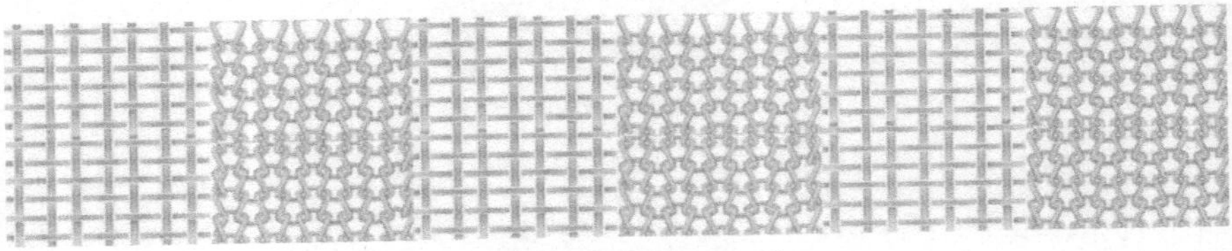

Length Wise Compounding

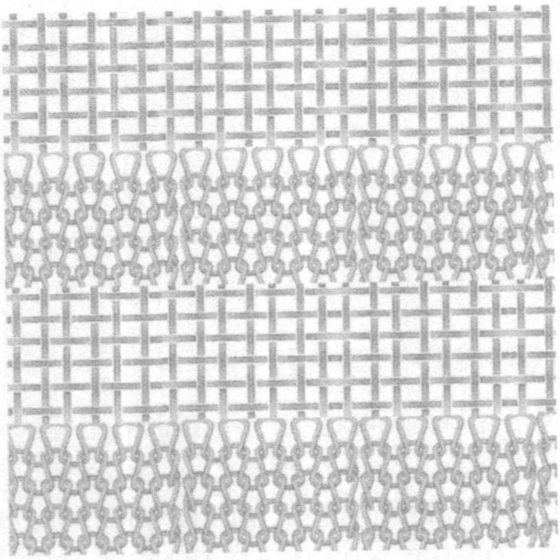

Width Wise Compounding

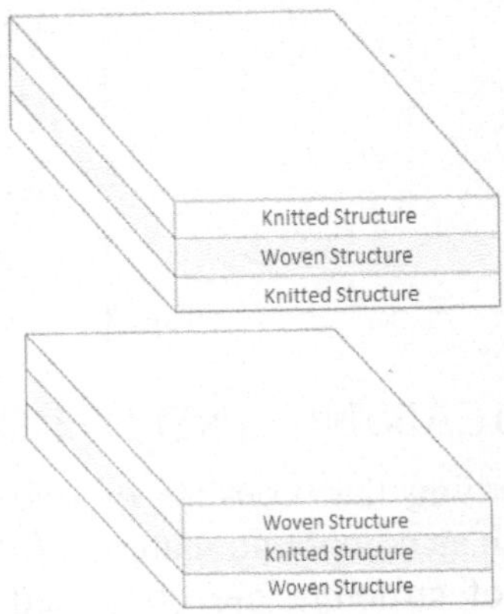

Layer Wise Compounding

Fig. 2: Different Ways of Fabric Compounding

3.3 Factors to Fix at Compounding

While compounding woven and knitted structure many things should be optimised like the seams and stitches at the joining part. The fabric strip length in case of length wise and width wise compounding and thickness in layer wise compounding. Also, the ticket number of thread, needle to be used should be optimized.

4. SPECIAL CARE TO BE TAKEN WHILE STITCHING WOVEN AND KNITS TOGETHER

A Seam can be defined as joint where a sequence of stitches unites two or more pieces of material. Seam must be balanced. It decides seam strength and also seam efficiency. while stitching these two-structure knitted and woven that has same seam length a special precaution must be taken at stretch as knitted fabric can get stretched and this may lead to defects. Seam puckering and waviness in the seam may occur due to stretching in the knit construction. So special pressure foot (walking pressure foot) should be chosen which will help to make seam stable. If walking foot cannot be attached to machine, then testing should be done at raw fabrics to check the stability of woven and knitted seam. Knitted fabric on top of the woven, or the other way around can be tried. One more difficulty is that, the knitted fabric may not have enough stability to hold the weight. Also, they may not drape without pulling unattractively at the seam. It can be corrected, by adding stabilization by adding elastic material with seam. The can be done by elastic that is the same width as the bodice. This extra elastic adds some structure to the seam. Woven fabric may fray while knitted may remain stable this precaution should be taken. So, any time you are putting these two structures together, one need to finish the seam after stitching them together or finish the edge of the woven fabric with pinking shears before stitching it to the knit.

5. CONCLUSION

Though woven and knitted structure has advantages at individual level, but its compounding may give better result and this compounding can be done in various ways like lengthwise, width wise and layer wise. The special study can be carried out on the joining of these structure.

REFERENCES

[1] Banerjee, P.K. (2014). Principles of Fabric Formation (1st ed.). CRC Press.

[2] Cook, J.G. (2001). Handbook of Textile Fibres Natural Fibers. Woodhead Publications Limited, 1, 3–8.

[3] Hayavadana, J. (2015). Woven fabric structure design and product planning (1st ed., Vol. 1), WPI Publishing.

[4] Ward, W.A. (2018). Beginner's guide to sewing with knitted fabrics: everything you need to know to make 20 essential garments (1st Ed.). CICO Books.

[5] Özdil, N., Marmarali, A. and Kretzschmar, S.D. (2007). "Effect of yarn properties on thermal comfort of knitted fabrics". International Journal of Thermal Sciences, 46(12), 1318–1322. https://doi.org/10.1016/J.IJTHERMALSCI.2006.12.002

[6] Choudhary, A.K. and Goel, A. (2013). "Effect of Some Fabric and Sewing Conditions on Apparel Seam Characteristics". Journal of Textiles, 1–7. https://doi.org/10.1155/2013/157034.

[7] Cooklin, G., Hayes, S. and McLoughlin, J. (2006). Introduction to Clothing Manufacture. In Introduction to Clothing Manufacture (2nd ed., Vol. 1). Blackwell Publishing. https://www.wiley.com/enus/Introduction+to+Clothing+Manufacture%2C+2nd+Edition-p-9780632058464.

[8] Wovens Knits Sewing them Together –Pollywoggles (2022) (n.d.). Retrieved January 20, from https://pollywoggles.ca/blog/wovens-knits-sewing-them-together

Contemporary Staple Yarn Spinning Systems

Rajesh T. Dhore and Shamayita Patra*
Shri Vaishnav Institute of Textile Technology, SVVV, Indore, India
E-mail: *shamayitapatra@svvv.edu.in

ABSTRACT: *A long continuous strand of fibers bound together is known as yarn. Yarns are classified on the basis of their composition and structure. They are filament, staple, and composite yarns. The skill of yarn spinning has begun approximately 8000 years ago. Actual mechanization of the spinning process took place in mid 1700s. Further developments were documented in the yarn spinning process up to 1960 with the invention of rotor spinning. Today the existing modern yarn spinning systems are Ring, Air jet, Friction, and Rotor spinning systems. All these processes produce yarn by inserting twists in the assembly of staple fibers to form a continuous structure. Out of all these spinning systems, ring spinning is the most versatile and widely used method. It covers around 80% of the global yarn production. In this review cumulative accounts are presented over their technical limitations and the properties of the yielded yarn structures.*

Keywords: Staple Yarn, Ring Spinning System, Air Jet Spinning System, Friction Spinning System, Rotor Spinning System.

1. INTRODUCTION

A linear assembly of fibres, in which fibres are held together, by imparting twist to form a long, continuous strand of fibres with small in cross-sections is known as staple yarn. This yarn is then used for fabric formation or any other process. To hold these fibres together some binding force is required which is provided by twisting fibres around yarn axis. Various natural or synthetic fibres are used to form yarn. The twist insertion in fibre assembly is carried out by various techniques. Initially, the twist is inserted in fibres assembly by rolling the attenuated length between the hands. Later on, drop spindle method is used for a long time for spinning yarns by using different fibres. The invention of the spinning wheel from 500–1000 AD [1] Figure 1. To form yarn from fibre assembly one of the inventions is the spinning wheel which used for yarn production till 1764 up to the invention of spinning jenny. The water frame was developed and patented by Arkwright in the 1770s. [1]

The actual ring-spinning frame was invented in 1832 by John Thorp. [2] The twist is inserted with the help of ring and traveller, whirling force in case of air-jet and fibre wrapping in rotor spinning.

Fig. 1: (a) Spinning Wheel, (b) Drop Spindle [2]

2. RING FRAME SPINNING

Ring spinning with the traveller is invented in the USA by Messrs Addison and Stevens in 1829. [3] The main limitation of ring frame is it consumes about 65–70% of total power requirement for driving spindle, 25% used for drafting of material and 5–10% lifer mechanism. [4] The range of spindle speed is from 6000–25000 revolutions per minute. It yields a maximum ring traveller speed of up to 45 m/s. The ring spinning system is adapted by both finer yarn count and courser yarn count staple fibre spinning. It is the most versatile machine for yarn production to produce a wide range of yarn counts from 0.5 Nec–120 Nec Figure 2.

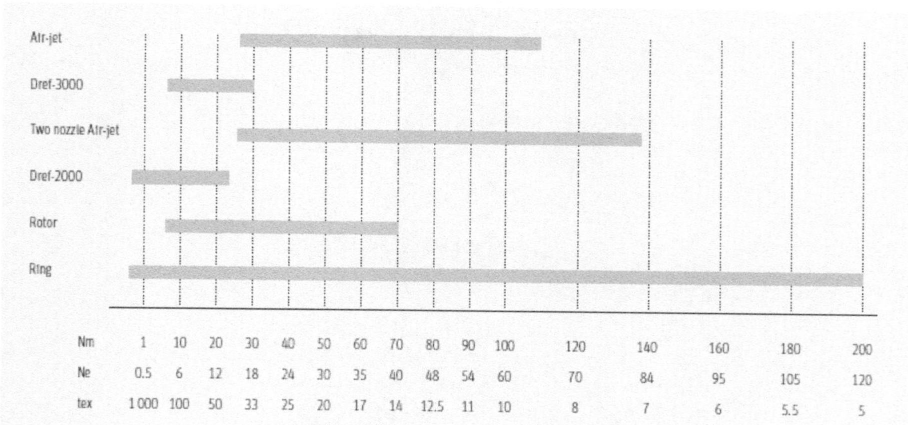

Fig. 2: Yarn Count Range of the Industrial Spinning System [5]

3. AIR JET SPINNING

In the year of 1960 DuPont (USA) was the first who bought the concept for air vertex or air-jet spinning. The patented method produced yarn called a 'fasciated yarn'.

In this spinning system twist inserted in fibres by vertex created with one or two air jet nozzles. The air-jet spun yarn has core parallel fibres wrapped with sheath fibres. An important feature of the process is the high rate of twist insertion which makes it possible to have high yarn delivery speeds compared with ring spinning, particularly with fine counts. [3]

The process involves formation of yarn from sliver there is no need of roving or speed frame. The machine is also equipped with piecing, auto doffing, yarn clearing, winding and package transfer operations. Murata has done most of the developments in the area of design for jet spinning in the early 1980s. These new design features commercially help to process cotton, short-staple man-made fibres, or blends for yarn counts ranging 20 Nec–80 Nec.

A new design for air-jet spinning can produce yarn at speed of 150–450 m/min. This is about 2–3 times rotor spinning and 20–30 times that of ring yarn spinning system. Due to core and sheath structure of yarn, only surface fibre contributes to as binding fibres, which results in low yarn strength. Also, the production of finer yarn is not possible in this method. [2] The typical passage of material for air-jet spinning is shown in Figure 3.

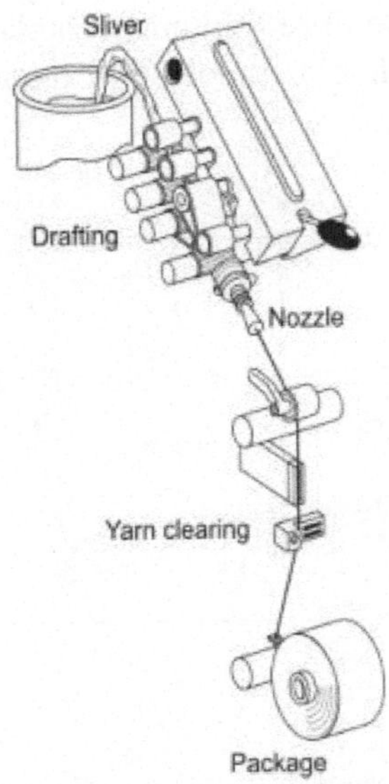

Fig. 3: Air Jet Spinning Principle by Murata [2]

4. FRICATION SPINNING

In 1973 DREF-1 friction spinning method was first developed and introduced by Dr. Fehrer A.G. from Austria. In friction spinning, twist inserted by frictional contact between the yarn surface and rotating cylinders or rollers (Figure 4). Latter in 1975, with some modification in design improved model of friction spinning DREF-2 has exhibited at ITMA exhibition. DREF-2 operates on the mechanical aerodynamic principle with internal suction arrangement through drums. It is useful to spun count in the range of 0.3–6 Nec.

DREF-3 is new improved version of DREF-2 which was introduced in 1981. The yarn quality is improved in this version and can spun count up to 18Nec. The yarn delivery rate for this system is about 300 m/min. It can produce core and sheath type of yarn structure.

Next to DREF-3 DREF-5 developed by Schalafhorst, Suessen and Fehrer. This system can spun count in the range of 16–40 Nec with a delivery speed of 200 m/min. But due to some reason, this model was not commercialized in the market. Latter 1999, in ITMA99 DREF2000 presented to the public as the latest development in friction spinning. This machine can spun count up to 14.5s Nec at a delivery speed of 250 m/min. Also in ITMA2003 DREF-3000 is presented as the next development for friction spinning which can spun count from 0.3 Nec to 14.5 Nec.

Some major limitations for friction spinning are, It has low yarn strength due to poor fiber orientation in yarn structure. The twist in yarn core and sheath has more difference. The system is also limited by drafting and fiber transportation speeds at various stages. [1-6]

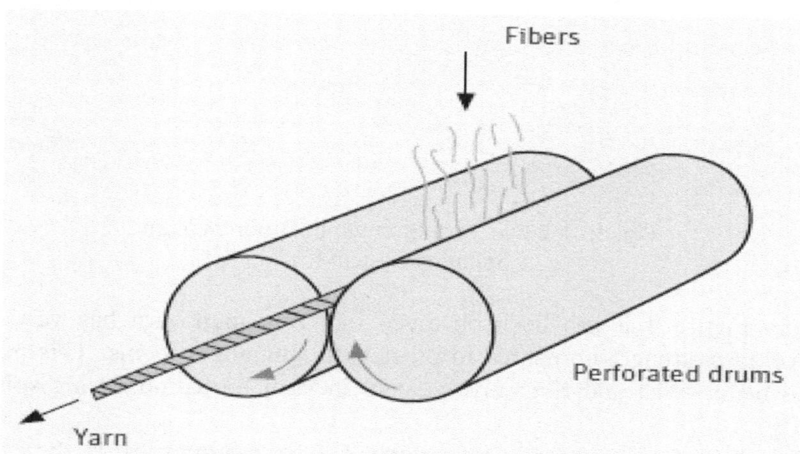

Fig. 4: Operating Principle of Friction Spinning [5]

5. ROTOR SPINNING/OPEN END SPINNING

As in 1960 huge increase in the production speed of yarn the preparatory machine the requirement of yarn production rate after draw frame motivated researchers for new inventions in yarn spinning. The idea for open-end spinning was developed in 1965 in Czechoslovakia. In this system, the yarn end is open for spinning process. Hence it is also known as rotor spinning or break spinning. In this process yarn spinning and winding operations are combined in one machine. So the winding operation is eliminated in this process. As rotor spinning combines two operations in a single machine i.e. yarn spinning and winding. Rotor spinning has more potential to twist insertion as the speed of rotor is about 150,000 rpm. Also only feeding carded sliver yarn production is possible in rotor spinning machine. It is more cost-effective in course count spinning when the rate of production is a concern. Rotor spinning is limited for course yarn spinning as it can spin count from 5–40 Nec.

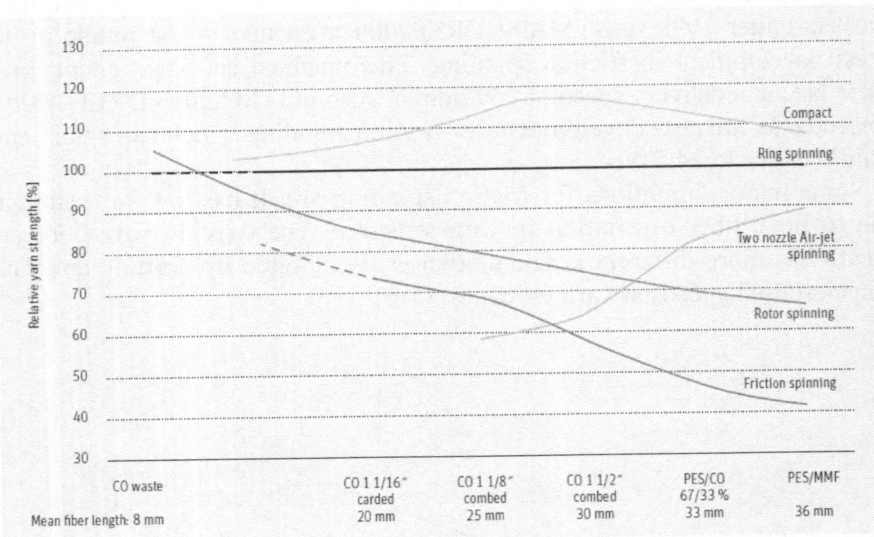

Fig. 4: Relative Yarn Strength in Different Yarn Spinning System [5]

From Figure 4 it has been observed that ring-spun yarn has very good relative yarn strength compared to other yarn spinning systems. This may be due to better twist and fiber involvement in yarn formation during spinning process.

Table 1 shows the order of importance for fiber properties. In the case of ring and air-jet spinning prime importance is given to fiber length. [1-6]

Table 1: Fiber Properties in the Different Spinning Systems [6]

Order of Importance	*Ring Spinning System*	*Rotor/Open-end Spinning System*	*Air Jet Spinning System*	*Friction Spinning System*
1.	Length	Fineness	Length	Strength
2.	Strength	Strength	Cleanliness	Fineness
3.	Fineness	Length	Fineness	Length
4.	Cleanliness	Cleanliness	Strength	Cleanliness

6. CONCLUSION

Above all yarn spinning system has its uniqueness. It has been observed that each spinning system has its advantage along with some constraints. Ring spinning system has more versatile compared to others, which can spin wide yarn count spinning range. In the count range of 60 Nec to 160 Nec ring frame still, dominate the market for cotton yarn production.

In case, of course, yarn count the rotor yarn spinning system proves more advantageous to all other systems. It has low labour, power, and machine requirement compared to other yarn spinning systems. Hence it proves cost, power, and labour effective spinning system for course yarn spinning. It has a big market share in course yarn count in the range of 6–20 Nec.

Air-jet spinning has replaced ring spinning potentially in the medium count range of 10 Nec to 60 Nec. It also consumes high energy costs as it requires compressed air.

In Friction spinning, the yarn production cost is less as compared to another spinning system. It also has a limited course count range to produce. The mechanism used to produce frication spun yarn is quite complex.

REFERENCES

[1] Williams, M. and Douglas, F. (1992). Cotton Mills of Greater Manchester. Carnegie Publishing.

[2] Lawrence, C.A. (2010). Advances in Yarn Spinning Technology, Wood head Publishing Limited.

[3] Oxtoby, E. (1987). Spun Yarn Technology, Butterworth Publication, 121.

[4] Werner, K. and Herbert, S. (2014). Ring spinning, *The Rieter Manual of Spinning* (Vol. 4).

[5] Herbert, S. (2014). Alternative Spinning System, *The Rieter Manual of Spinning* (Vol. 6).

[6] Heinz, E. (2014). Rotor Spinning, *The Rieter Manual of Spinning* (Vol. 5).

Application of Braiding Technology for Development of New Jute Technical Textile Products

Mahuya Ghosh
Scientist, Indian Jute Industries' Research association (IJIRA), Kolkata–700088, India
E-mail: mahuya.g@gmail.com

ABSTRACT: *The concept and scope of two-dimensional circular braiding technology for development of different new jute technical textiles are described in this paper emphasizing on development of a particular product, namely, jute braided sapling bag. A tri-axial braided construction was chosen to develop nursery sapling bag from jute by employing a novel process. This product is meant as an eco-friendly substitute of plastic-based nursery sapling polybags which are very commonly used in agri-horticultural or forest nurseries for raising and transporting plant seedlings. To develop a cost-effective product with high fabric cover and low mass, the scope of standard jute yarns (fine and coarse both), jute rove and 'jute tape' was explored as 'axial material' during circular braiding. 'Jute tape' was found to be effective in providing better cover and lower mass in the product. The developed jute tape-based sapling bag was found to be superior to polybags in respect of soil-loss during handling in nursery. Additionally, soil-analysis test revealed that bio-degraded jute bag enhanced soil nutrient level considerably with organic carbon, potassium and phosphorus.*

1. INTRODUCTION

Jute is an expensive mechanically robust, hygroscopic and bio-degradable natural fibre. It is abundance in our India. These attributes render jute as highly suitable for uses in the field of agrotextiles, geotextiles and other emerging technical textiles in addition to its traditional packaging application.

Braiding is a simple but versatile single-stage narrow fabric manufacture technology. There are several types of braiding processes, e.g., flat braiding, circular braiding, tri-axial circular braiding and three-dimensional braiding (used mainly for preparation of composites). The first three types of braiding are two-dimensional based on construction and also used in composite industry. In this paper, discussion will be restricted in to relatively simpler two-dimensional braiding technology. Two-dimensional braiding has some advantages over conventional weaving technology. Such as,

- This technology does not require warp and weft yarn preparatory processes, e.g., warping, sizing, beaming, preparation of pirns/cops etc. as needed for weaving.
- Tri-axial circular braiding process would produce tubular fabric or sleeve directly from yarn having some extra cover and reinforcement along the sleeve axis than braided sleeve without axial material.
- Elimination of Side-stitching for circular braided product,
- Wide range of raw material variety, viz., yarns/tapes/roves/wires
- Lesser number of machinery requirements reduces the production cost.

There are some existing braided products, viz., jute ropes (Figure 1), mats, bag handle, handicraft items, etc. in India and abroad. Most of these products are hand braided adopting flat braiding technique. The fully mechanized circular braiding technology has capability to produce a varied range of products with respect to cost and application areas. But, Jute Industry is yet to adopt this technology for product diversification.

Fig. 1: Jute Flat Braided

Sapling bags are designed to foster plant saplings in horticulture, forestry and agro-forestry nurseries as well as permit their efficient transport to fields. Such products fall under the category of technical textiles and serve simultaneously functions of Agrotextile and Packtextile.

Nursery sapling bags enjoy a considerable market demand evidenced for example by the requirement of 2000 million pieces in year 2012 as predicted by the working group on horticulture and plantation crops for eleventh five year plan of Government of India. Similarly for greening of Sikkim, the target fixed by the government was to raise 100 million quality planting material (QPM) seedlings beginning from the year 2009–2010. [1] These bags can be useful for raising and transporting costly ornamental saplings and cactuses (xerophytes). Discarded polybags after transplantation of saplings in the field pose serious environmental hazards unless suitable recycling facilities of these waste materials are in place. Equivalent jute sapling bags should however pose no such problems and evidently can even be planted in field along with the saplings for providing nutrition to soil during their natural process of biodegradation. Moreover during the period of nurturing of saplings in nurseries, jute sapling bags can be expected to provide superior microclimate for the plant roots for the natural porosity of jute fibres and the fabric themselves. Such bags would be able to contain wet clayey soil efficiently. Multistages manufacturing process increases final cost of the jute sapling bags.

A very few standard jute mills produce this type of woven sapling bags of different sizes on order basis from abroad. Considering the unique features of circular braiding technology, a project was taken up to explore the concept of developing relatively cost-effective jute braided sapling bag and related issues at the Indian Jute Industries Research Association (IJIRA). The major outcome of which is reported in this paper. [2–4]

2. MACHINERY, MATERIALS AND METHODOLOGY

After a meticulous review of related works [5,6] it was decided to develop a braided jute sapling bag of dimension, 230 mm (length) × 150 mm (folded width) having maximum possible fabric cover with minimum mass. Accordingly a tri-axial circular jute braiding machine and winding unit were designed and fabricated by IJIRA (Figure 2). The braiding machine has 48-spindles having provision for inserting 24-axial materials. The fabric take-up speed and braiding speed are independently controllable with the aid of Programmable Logic Control (PLC) system. The machine is capable of producing braided sleeves with maximum folded width of 150 mm while sleeves of lower width can also be easily prepared. The resultant product would be a continuous tri-axial braided sleeve. Tri-axial braiding is done by introducing a third set of yarns/tapes/roves (axial materials) parallel to the braid axis during braiding. These axial yarns do not take part in the braiding process but only get entrapped in the braided structure thus remaining perfectly aligned along the braid axis imparting a level of rigidity and cover to the fabric. [7] Further modifications were incorporated in the braiding machine as and when necessary to develop the desired product.

To make the product cost-effective along with maximum possible fabric cover and low mass simultaneously, a series of trials and systematic evaluation of products were carried out at IJIRA Pilot Plant. It may be noted that sapling bags are not as such subjected to any mechanical strain during their end-use and hence the role of the covering material is simply protective in nature i.e. of prevention of soil loss and of providing a moist and favourable environment for healthy root growth. Clearly slivers by their inherent nature should provide greater cover than resultant twisted yarns, and can retain more moisture in their inter fiber spaces. But, to arrest the disintegration of slivers on the braiding machine during withdrawal process from feed zone into braiding zone, some strengthening through application of suitable adhesives was adopted employing a suitably modified padding mangle. In this way, finally, a novel material, named as, '***jute tape***' (adhesive bonded flattened jute sliver) of suitable dimension and strength, was created and subsequently introduced as the axial material during braiding. Simultaneously high fabric cover and low mass in the product. A *tri-axially braided sleeve, an axial assemblage of several 'jute tapes' bound in position with minimum of number of braiding yarns produced.* Evidently, the 'jute tape' was prepared from very light weight jute slivers from medium or lower batch quality. Performance is given to thinner tapes with ideal amount of structural strength. The braided sleeves prepared from 2102 tex jute slivers (i.e., axial materials) and 275.7 tex jute yarns (i.e. braiding yarns) were reasonably thin with almost half the mass per unit area than that of the sleeves produced with jute roves. The jute tape based braided sleeve is shown in Figure 3.

Fig. 2: The Tri-Axial Jute Braiding Machine Developed by IJIRA

Fig. 3: Jute Braided Sleeve Composed of 24 Jute Tapes and 48 × 2 Jute Yarns

The continuous braided tubes were cut into pieces and the individual pieces converted into sapling bags by doubling the above-mentioned jute tape-based sleeves in a specific manner and closing one end suitably. Actual photographs of this bag product and of such bagsplanted with saplings are shown in Fig.4 and 5 respectively.

Fig. 4: Jute Braided Sapling Bag

Fig. 5: Planted Jute Braided Sapling Bags in Nursery

3. THE PRODUCT AND ITS PERFORMANCE EVALUATION

The jute braided sapling bag selected for performance evaluation had the following specifications:

1. Mass: 40 g (approx.)
2. Dimension: 230 mm (length) × 130 mm (folded width)
3. Construction: 2/2 regular tri-axial braid employing 48 braiding strands each containing two yarns each of count 275.7 texwhile the axial jute tape was prepared from 2102 texjute sliver.

Actual field tests of these bags were carried out for evaluating their performance in taking care of nursery saplings (Sunflower saplings) during their stay period in nursery and their role even after transplantation in field. Performance evaluation of this agrotextile product was done on the basis of few morphological seeding traits, viz., stem base diameter (SBD), stem height (SH), root volume and length, etc. A commercial polybag was used as the control.

Minimal SBD growth was observed for saplings planted in jute braided sapling bags than those planted in polybags while reverse is true with respect to SH growth. But statistical analysis (t-test) reveals that these differences are not significant at 95% and 99% levels of confidence. Hence, it can be concluded that growth of sunflower seedlings with respect to the morphological traits, viz., SBD and SH results similar or comparable in both types of sapling bag under experimentation.

It was observed that emergence of a healthy root network through the jute braided sapling bag while a single root is able to come out through the hole punched in polybag. This is due to the fact that jute braided sapling bags are porous while porosity of polybags is almost zero. Consequently, due to better air circulation inside the soil mass of jute sapling bags healthy root networking was observed.

Soil test was carried out on control soil samples and treated soil samples (soil with degraded jute) after the lapse of a predetermined time (in this case 55 days after transplantation of sapling with jute sapling bag in field) to document nutritional level of soil.

It was very clear that bio degraded jute enhanced nutrient level of soil detail data are given in (Table 1).

Table 1: Soil Analysis Result.

Sample Type	*pH*	*Organic Carbon (%)*	*N (kg/ha)*	*P2O5 (kg/ha)*	*K2O (kg/ha)*
Control soil	7.7	0.79	384.0	66.3	163.7
Treated soil	7.5	1.50	361.0	118.5	371.5

In a horticulture nursery, soil loss from jute braided sapling bags and nursery sapling polybags (with air circulation and water drainage openings) of similar dimension was assessed. During this period (until the soil loss from jute bags becomes marginally low, here 10 days) the experimental sapling bags planted with saplings were kept on plastic trays to collect drained out soil.

Initially elevated and gradually decreasing soil loss was observed for the jute sapling. After 6 days it was negligible. Whereas, very slow soil loss reduction trend was observed for polybag counterpart (Table2).

Table 2: Soil Loss from Sapling Bags in Horticultural Nursery

Soil Collected on	*Soil Loss from 15 Planted Jute Braided Sapling Bags (g)*	*Soil Loss Per Jute Braided Sapling Bag (g)*	*Soil Loss from 15 Planted Polybags (g)*	*Soil Loss per Polybag (g)*
Third day	16.05	1.06 (in 2 days)	13.90	0.93 (in 2 days)
Ninth day	5.30	0.35 (in 6 days)	14.50	0.97 (in 6 days)

4. CONCLUSION

Jute tape, prepared from jute sliver of 2102 tex, was used as axial material to develop the desired low mass and high cover jute braided sapling bag product. Jute tape based sleeve is having 0.58 times the mass per unit area in comparison to that of all coarse jute yarn based sleeve while width and fabric cover for those are comparable.

Actual field trials employing jute braided sapling bag and polybag reveal that:

- Biodegraded jute bag enhances soil-nutrient level by enriching soil considerably with organic carbon, phosphorus and potassium.
- Jute braided sapling bag allows healthy root network growth and unrestricted emergence of roots through its openings.
- Growth of sunflower seedlings with respect to SBD and SH are similar or comparable in both type of sapling bags.
- Performance of jute braided bag is far better than polybags in respect of soil loss

Hence, it can be concluded that jute braided sapling bag can advantageously replace synthetic eco-hazardous polybags. To compare costing of the two products, suitable scaling up of the production process ofthe jute braided sapling bag is essential and ecological costs should be given due consideration.

The braided jute sleeves so produced on the braiding machine can in principle be employed for protection, storage and transportation of natural materials. The developed nursery sapling bag is one such product. With suitable manipulation one can create structures for storage and transport of fruits and vegetables as also Geotextiles (viz, Pre-fabricated vertical drain for soft soil consolidation, encasement of grout, etc.), other Agrotextiles (e.g., wrapper fabric sleeve for Moss-sticks used in home gardens) etc. A research work was already carried out by Sampath Kumar to develop a prefabricated jute-coir braided drain product successfully at laboratory scale at Indian Institute of Technology Delhi. [7] Additionally, the same research team from IJIRA related to development of jute braided sapling bag has already ventured into the development of suitable jute braided sleeve for Moss- stick pre-parathion. [2] In fact the developed braiding technology has potential to open up an entirely new prospect of jute diversified products.

ACKNOWLEDGEMENT

The author is sincerely grateful to National Jute Board (NJB), a statutory body under Ministry of Textiles, Govt. of India, for necessary funding to carry out the research project under Jute Technology Mission scheme. The author is highly obliged to Dr. Prabir Ray, the then Director, IJIRA for his support and Dr.-Ing. (Prof.) P.K. Banerjee, Project Advisor for his guidance.

REFERENCES

[1] Lachungpa, C. The Status of Forest Nurseries and Future Action Plan 2009–2010. *Forests, Environment and Wildlife Management Department*, Government of Sikkim, 2009.http://www.sikkimforest.gov.in/docs/Forestry/Status%20of%20Forest%20Nurseries%20and%20Furture%20Action%20Plan%202009-10.pdf.

[2] Project Completion Report on “Manufacture of Jute Braided Cloth by Appropriate Design Incorporation in Braiding Machines”, Jute Technology Mission (Mini-Mission IV, DDS 7.1), Project No. 4, sponsored by National Jute Board—Astatutory body of Ministry of Textiles, Govt. of India, 2013.

[3] Ghosh, M., Biswas, D.K. and Sanyal, P. (2016). “Development of jute braided sapling bag for nursery use”. *Journal of Natural Fibers*, 13(2), 146–157.

[4] Banerjee, P.K. and Ghosh, M. (2014). “Development of Braided Jute Sapling Bag”. *International Conference on Technical Textiles and Nonwovens-2014*, Indian Institute of Technology Delhi, November, New Delhi, India.

[5] Howell, K.D. (2004). “Nursery practices influence seedling morphology, field performance, and cost efficiency of containerized Cherrybark oak”. *Southern Journal of Applied Forestry*, 28 (3), 152–162.

[6] Ferdousee, N., Jabbar, F., Hossain, M.K. and Hoque, R.A.T.M. (2010). “Comparative Growth performance of Leucocephala seedlings raised in nursery bed, polybag and root trainers”. *Proceedings of International Conference on Environmental Aspects of Bangladesh (ICEAB10),* 65–68.

[7] Jinka, S.K.P. (2000). “Studies on Braided Prefabricated Vertical Drain”. *Indian Institute of Technology Delhi, India.*

Self Cleaning Finishes on Textile Fabrics

S. Chinnamma,* M. Indhujaa, A. Arokiya Reena, Maria Rubin Aswini. A, M.R. Karthikeyan and R. Gobinath

Department of Fashion Technology, Sona College of Technology, Salem–636004, Tamil Nadu, India

E-mail: *23102002chinnu@gmail.com

ABSTRACT: *The self-cleaning idea can be applied to a variety of materials, including fabrics that are often used in daily life. This technology can also be used in other fields, including medicinal textiles, sporting clothing, military uniforms, and outdoor materials. It is also advantageous since it successfully conserves water and improves the beauty of the surroundings, and it will save electricity, laundry costs, and time in the long run.In today's world, there are numerous inventions available. Self-cleaning technology is one of them, and it comes from nature. Butterfly wings and plant leaves such as cabbage and lotus are just a few examples. In order to impart high-performance capabilities on natural fibres, natural fibre absorption and/or penetration of functionality of polymeric materials on/into fibre surfaces will be helpful. Because of the vast range of applications, such as window washing, solar panels, and cement, this technology has received a lot of attention.*

Keywords: Self-cleaning, Medical Textiles, Natural Fibres, Functional Polymer.

1. INTRODUCTION

Textile scientists and technicians are developing a plethora of novel textile materials and fabrics. It would be better if Clothing could stay clean for a few days without needing to be washed. This would significantly reduce the expense of cleaning the fabric or garment while also extending the textile's lifespan. For years, numerous textile innovators have struggled to keep water and dirt out of their fabrics. To create self-cleaning cloths, researchers used nanotechnology and versatile chemical treatments. Biomimicry in textiles has led to the development of water and dirt resistant treatments. The concept is based on the wicking capabilities of lotus leaves, which help to keep the surface clean. A fabric has been developed using nanotechnology that not only keeps the exterior clean but also keeps the textile material polluted and bacteria-free. [1-5]

2. DESCRIPTION

Self-cleaning coatings are divided into two categories: hydrophilic and hydrophobic coats. A hydrophobic coating with high water contact angles, above 90 degrees, is used to create the lotus effect. The contact angle is calculated with the help of the surface's chemical and topographical properties. The water contact angle of a superhydrophobic covering is larger than 150°. The dirt is taken away by the spherical water droplets.

The nano particles utilized in the coating have a high surface energy and a big surface area to volume ratio. As a result, particles have a better affinity for fabrics, allowing the functional coating to remain longer. A lot of research and advancement in the nano technology application are documented. There are materials available today that can repel not only water and grime, but also wine and coffee stains, bacteria, and stink.

Using photocatalysts, a mixture of titanium dioxide and zinc oxide that acts as a self-cleaning and anti-bacterial agent, is one of the approaches for coating nano particles. Clothes are coated with a thin layer of 20mm titanium dioxide particles. When exposed to light, the surface of the textile material breaks down contaminants like dirt, pollutants, and microorganisms that come into touch with it into carbon dioxide and oxygen.

It removes all types of stains and contaminants from the surface.

Microwaves are used in another technique. The surface is coated with a chemical that repels dirt and oil, and nano particles are employed to create a protective layer. Defense troops utilise uniforms and innerwear made with this technology. These coatings have the capacity to keep microorganisms at bay while keeping the cloth clean for days.

Silver also has a natural ability to repel dirt and microorganisms. Silver injected nano particles to create self-cleaning fabrics. Silver is also thought to have healing properties, which is why it is utilized in items like stockings. Chlorine is another chemical that has self-cleaning characteristics. It is employed in the form of halamine molecules, which, unlike chlorine in gas form, have no negative consequences. Chlorine halamines are bonded to cotton cellulose fibres, and the fabric is then treated with chlorine bleach to restore its cleaning capabilities after a few washes. Such a treatment is employed in medical textiles including wipes, beddings, and towels.

Self-cleaning clothing and fabrics maintain the fabric's original texture and feel. They also save a lot of water and energy in terms of less offer washing. Self-cleaning fabrics can avoid filth, oil, and act as a disinfectant when using the technologies stated above. The processes utilized to create self-cleaning fabrics ensure that the textiles endure longer and look better than standard fabrics. [1-5]

3. APPLICATIONS

Anti-dirt and self-cleaning clothing can be found in a variety of applications including medical, sports, defense, and household textiles. Some of the procedures used, however, have flaws. The amount of filth or stains to be cleansed determines how long it takes to break down pollutants. Titanium dioxide contains electrons that use light energy to react with oxygen. A huge stain on a fabric will need to be exposed to sunshine for a longer period of time in order to entirely break down. In addition, producing hydrophobic textile materials takes time and money. [1-5]

4. CONCLUSION

Refining the technology for self-cleaning fabrics and clothes will assist reduce costs while also ensuring that the nanoparticles utilized have a long-lasting effect. This green option can help you save money on dry cleaning and laundry expenses, as well as energy and time. In the not-too-distant future, clothes with self-cleaning capabilities will be accessible commercially in stores.

REFERENCES

[1] "Self-cleaning textiles". *Fibre2fashion.com.* https://www.google.com/search?q=1.+Self-cleaning+textilesBy+Fibre2fashion.com&oq=1.%09Selfcleaning+textilesBy+Fibre2fashion.com&aqs=chrome..69i57.1581j0j15&sourceid=chrome&ie=UTF-8.

[2] Malik, T. "Self Cleaning Textile—An Overview". *TechnicalTextile.net*, https://www.google.com/search?q=Technical+Textile.net&ei=eaj7Ycy3EIWS-AbIgqrYAg&ved=0ahUKEwiM2NH7oP1AhUFCd4KHUiBCisQ4dUDCA4&uact=5&oq=Technical+Textile.net&gs_lcp=Cgdnd3Mtd2l6EAMyAggAMgIIAEoECEEYAEoECEYYAFAAWABg8xFoAHAAeACAAboDiAG6A5IBAzQtMZgBAKABAqABAcABAQ&sclient=gws-wiz.

[3] Afroz, S., Azady, A.M., Yeasmin, A., Abdullah, A.R., Zahid, H, Rahaman, M.S. and Islam, J.M.M. (2021). "Self-cleaning textiles: structure, fabrication and applications". The Textile Institute Book Series, Fundamentals of Natural Fibres and Textiles, Woodhead Publishing, 557–597.

[4] Saad, S.R., Mahmed, N., Mustafa, M.M.A.B. and Sandu, A.V. (2016). "Self-Cleaning Technology in Fabric: A Review". *IOP Conference Series: Material Science and Engineering,* 133 012028.

[5] Diaa, M. and Ahmed, G.H. (2022). "Self-Cleaning Properties of Cellulosic Fabrics—A Review". *Bio interface Research in Applied Chemistry*, 12(2), 1847–1855. https://doi.org/10.33263/BRIAC122.18471855

Multifunctional Finishing of Jute and Jute-Cotton Union Fabrics with Nano-Titania

Sandip Bose* **and Amal Chandra Deka**

Indian Jute Industries' Research Association, Kolkata, India

E-mail: *sandipbosu@gmail.com

ABSTRACT: *Cellulosic textile material exhibits many interesting properties when oxides of ceramic materials in nano-form are deposited on it. A method for application of nano-finished formulations on ligno-cellulosic textile like jute fabrics has been developed based sol-gel method. By pad-dry-cure process, at temperature exceeding 130°C, nano-titania coated ligno cellulosic textile could be developed. The resulting nano-material coated jute and jute-cotton union fabrics exhibit properties like hydrophobicity, anti-microbial property and UV-resistance. The nano-finished fabrics have potential of opening new avenues of application of cellulosic and blended textiles.*

Keywords: Nanofinishing, Titania, Sol-gel, Cellulosic Textiles, Functional Properties.

1. INTRODUCTION

Enhancement of desired functional properties of a textile material for definite end-uses, could be achieved by use of suitable textile finishing. Textile finishing has witnessed the advancement of finishing technologies like water repellence in textiles: from application of rubber tree gum and drying it over smoke fire by people of ancient Mayan civilization, to perfluorooctane sulphonate (PFOS) perfluorooctanoic acid (PFOA) and perfluorohexanoic acid (PFHA) based breathable finishing of textiles. In present scenario tofluorocarbon free finishing technologies are ginning impetus. The microorganisms present in the environment consume the cellulosic textile materials, like cotton and jute, as source of food. So, another important aspect of textile processing, especially for the cellulosic textiles, are anti-microbial finishes, that prevents the attack by microorganisms, and slows down the process of biodegradation of the textile materials. Many of the preservative chemicals used in the past, like copper based organo-metallic compounds, mercury and tin compounds, zinc, zirconium and cadmium-based compounds, lead and other heavy metals, phenols and related compounds and quaternary ammonium compounds have gradually been discontinued due to the environment and safety concerns. By correlating reports of skin cancerandprolonged exposure to sunlight which contains ultraviolet (UV) rays, inferences could be drawn to the dangers of UV exposure. Excessive exposure to sunlight has been attributedas cause of damage to skin (like erythema, certain skin cancers) and eyes (like keratitis and cataracts). UV-protective textiles, UV-protection sunglasses, and sunscreen creams are protective tools to safeguard against UV-radiation.

Matteris made up of molecules. The way in which these molecules are oriented, especially at the surface, determines the surface properties of these matter. By application of nanotechnology the structure of matter at molecular level is controlled in such a way so as to produce materials with strikingly different properties. The finishing of textiles with nano scale indentation on the surface can impart multifunctional properties with a single material chemistry. Also, at a small quantity of add-on the finishing chemicals can be very effective as it is in finely dispersed state and it resides mostly at the surface. Silver nano particles imparting anti-microbial property, oxides of zinc and titanium improving UV-protection property of textiles and silica-based nano particles improving hydrophobic character are some examples of promising use of nano particles on textiles. The advantage of these finishes is low add-on (of the order of 10 ppm to 1000 ppm) thus economizing the cost of finishing. The sol-gel technique has acquired considerable importance in synthesis of the nano particles due to its ease of application and low temperature phase formation.

2. MATERIALS AND METHODS

Titanium is opropoxide (TTIP) (97%) (Sigma Aldrich), was used along with industrial grade (Emplura®) acetic acid and 2-propanol (Merck) for preparation of titanium dioxide. Jute fabric of 458 g/m^{2}andJute cotton union fabric of 315 g/m^2 (ends/dm × picks/dm: 262 × 112) were used.

TTIP was diluted with 2-propanol (1:1) and acetic acid was slowly added to the TTIP-alcohol mixture while stirring the mixture preferably at 0–10°C. It is necessary to control the exothermic reaction during sol formation to prevent excessively large particle formation. It has been observed TTIP exchanges isopropyl groups with acetate groups. It results in molecules with approximate formula of Ti $(OPr)_2$ $(OAc)_2$. The sol could be diluted with water and added to the trough of a padding mangle where the jute fabric was impregnated. The fabrics was dried at 105°C and cured at 140–150°C, with residence time of 90–120 seconds, to obtain the finished fabric. Theadd-on of application could be varied in the range of 10–1000 ppm by controlling the dilution of the sol. [1, 2]

The finished jute and jute cotton union fabrics were assessed as per with AATCC Test method 100–2004 for antimicrobial study. The studies were carried out with *Staphylococcus aureus* and *Klebesiella pneumoniae*.

The hydrophobic property was estimated by observing the water contact angle aqueous liquid repellency rating. The fabric was supported on an embroidery hoop. A droplet of approximately 0.04 ml was dropped by a pipette onto the fabric from a height of 0.6 cm. The contact angle of the fabric surface was estimated using macro photography. The camera was positioned on a tripod, so that, the centre of the lens and the top surface of the fabric are located on the same plane. The aqueous liquid repellency rating of the finished jute and jute-cotton fabric were tested by AATCC 193–2005.

The UV protection capability of the nano titania finished fabric was assessed as per the AATCC-183 (2004) test method using UV-vis spectrophotometer (Varian Cary, 5000). The Varian, Cary 5000, with spectral range of 175–3, 300 nm was used to assess the UPF. The UV transmission and reflection profiles of the control samples in comparison to the treated sample spectra and the efficacy of shielding the UV radiation was evaluated. The average of four scans obtained by rotating the sample by 90° was reported.

3. RESULTS AND DISCUSSIONS

The results of antibacterial study of titania finished jute-cotton fabrics are narrated in Table 1.

Table 1: Antibacterial Study on Titania Finished Jute-Cotton Fabric with AATCC 100–2004

Sl. No.	*Sample*	*Test Organism*	*No. of Bacteria (CFU*)*		*Percentage Reduction*
			0 hours	*24 hours*	
1.	Jute-Cotton (Control)	*S. aureus*	1.4×10^5	1.25×10^7	Nil
		K. pneumoniae	1.86×10^5	1.46×10^7	Nil
2.	Jute-Cotton (300 ppm TiO_2)	*S. aureus*	1.01×10^5	< 10	99.99%
		K. pneumoniae	1.66×10^5	< 10	99.99%

When titaniais irradiated by light of sufficient energy, the photon excitation generates electron–hole pairs. It induces redox reactions at the surface of the coating. Consequently, electrons in the valence band of TiO_2 jumpsto the conduction band, and the electron and hole pairs are formed on the surface. Finally various active radicles that are formed by photoinitiated electrons and holes destroys the organic compounds of bacterial cell.

The hydrophobic nature of the fabrics assessed by contact angle measurements. This was found out from the water contact angle (WCA) which is shown in Figures 1 and 2 for Jute-cotton and jute fabrics respectively.

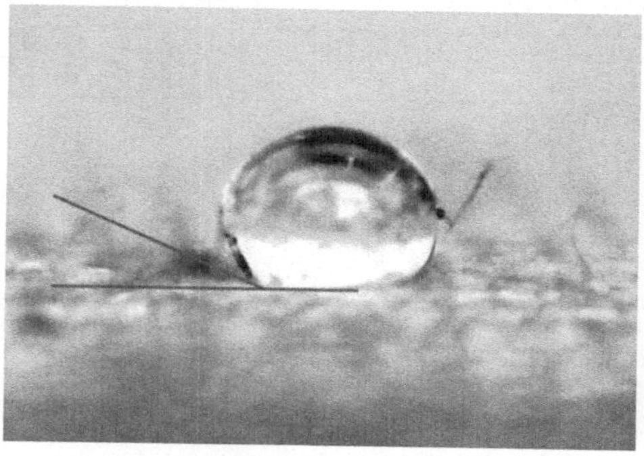

Fig. 1: Jute-Cotton Fabric Showing Contact Angle of 156° Treated with 300 ppm Titania

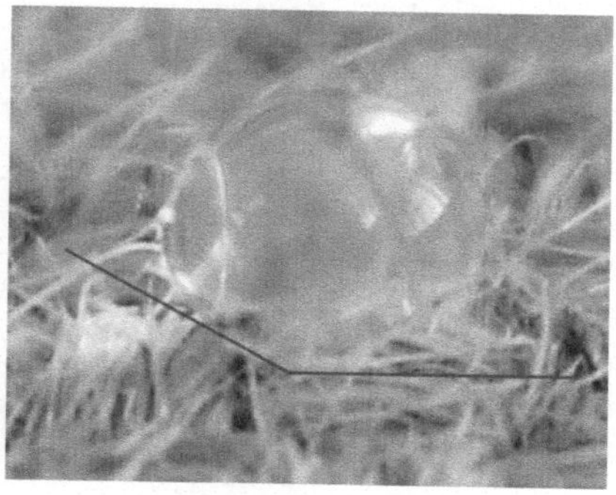

Fig. 2: Jute Fabric Finished with 100 ppm Titania Showing Contact Angle of 150°

The aqueous liquid repellency rating (AATCC 193–2005) of the finished jute and jute-cotton fabric is shown in Table 2. From the data it can be seen the water-alcohol solution repellence observed in the trial is better for Jute than Juco fabrics.

Table 2: Aqueous Liquid Repellency Rating of Titania Finished Cotton Fabrics in Industrial Trial by AATCC 193–2005 Test Method

Serial No.	*Sample Description*	*Aqueous Repellency Rating*
1.	Jute	4.0
2.	Jute-Cotton Union Fabric (315 gsm)	3.5

UV-spectro photometry method was used for determining UV protection factor of the jute and jute-cotton blended textiles. The UV registance of the fabric sample were measured by using AATCC- 183–2004. Results are represented in Table 3. The ratio of the average effective UV irradiance for unprotected skin to the skin protected by the test fabric is termed on ultra violet protection factor (UPF). The UPF depends on the types of composition fibre, yarn and fabric and on applied chemical finish.

From Table 3, it can be concluded that the UPF improved progressively with increase concentration of the finishing chemical. It noteworthy that titania, even at 100 ppm level of application, improves the UPF rating of the fabric significantly. From the tabulated data it may be concluded that the application of titania, irrespective of the level of application, improved the UPF rating of the treated fabrics. However, a higher loading provides a better protection.

Table 3: UPF of Titania Treated Jute Fabric by AATCC-183

Sl. No.	*Sample Description*	*Ultraviolet Protection Factor (UPF)*	
		UPF (Mean)	*UPF Rating*
1.	Jute Control	57.64	20
2.	Jute 100 ppm TiO_2	67.1	35
3.	Jute 1000 ppm TiO_2	140.83	45

Due to the high specific surface area nona scale titania remains adhered to the fibres. The TiO_2 work more effectively as it is present on the surface in finely dispersed state.

4. CONCLUSION

In order to improve the performance properties of textile materials ceramic materials were deposited on a variety of jute fabrics. These ceramic materials are capable of imparting multiple functional properties on textile substrate when deposited in nano form. TiO_2 had imparted the desired properties. A novel aqueous based titania application method was developed. Titania nano finished fabrics were found to impart hydrophobic property, UV-resistant property and anti-bacterial property to cellulosic textile materials like jute and jute-cotton fabrics.

Significant improvement in UV protective property (manifested by the UPF rating) was observed on titania treated samples compared to control fabric samples. Hydrophilicity studies showed more than 150°water contact angle on jute and jute-cotton fabrics at 100 ppm level of titania application. An aqueous liquid repellency rating of 3.5 was obtained with 100 ppmtitania finished jute-cotton, while a rating of 4on jute fabric was obtained by AATCC 193–2005. Titania finish did not deteriorated in tensile strength of fabrics. The shelf lives of the nano titania based formulations developed for nano-finishing were found to be reasonably good when stored in sealed containers.

5. ACKNOWLEDGEMENT

The authors acknowledges the financial support provided by Office of The Textile Commissioner, Mumbai, Ministry of Textiles, Government of India for carrying out the research on “Multifunctional Ceramic based Nano finishing of Outdoor Textiles by Sol-Gel Technique” at Indian Jute Industries’ Research Association (IJIRA).

REFERENCES

[1] Birnie, D.P. and Bendzkob, N.J. (1999). "^{1}H and ^{13}C NMR observation of the reaction of acetic acid with titanium isopropoxide". *Materials Chemistry and Physics,* 59(1), 26–35.

[2] Parra, R., Goes, M. and Castro, M.M.S. and Longo, E. (2008). "Reaction Pathway to the Synthesis of Anatase via the Chemical Modification of Titanium Isopropoxide with Acetic Acid". *Chemistry of Materials,* 20(1), 143–150.

Surgical Gowns for Healthcare and Medical Applications

V. Dharunya and S. Viju*
Department of Textile Technology, PSG College of Technology,
Coimbatore, Tamil Nadu, India
E-mail: *vijutext@gmail.com

ABSTRACT: *Medical textiles has emerged from the combination of textiles and their use in medical sciences. With the development of novel fibers and production processes for yarns and fabrics, new areas of application for medical textiles have been found. Development in the realm of textiles is usually targeted at improving user comfort. In health-care settings, gowns are an example of personal protection equipment. Gowns are used to limit the risk of transmission of infectious among healthcare workers, to reduce the risk of physician transmission, and to aid in the preservation of a sterile environment during invasive procedures. Surgical team members and nurses wear gowns in the operating room to avoid the transfer of germs and body fluids from the surgical team to the patient, as well as the patient to the crew. Surgical gowns are intended to handle a variety of procedures. Surgical gowns come in a variety of fabrics and fibres to protect surgical employees from infection and reduce microbial contamination of the incision. These gowns are either "disposable" or "reusable/multi-use/multiple usage." The design and performance characteristics of each of these categories varied significantly. This paper discusses about the functional requirements, material characteristics and design aspects of surgical gowns.*

Keywords: Surgical Gown, Design, Requirement, Types, Characteristics.

1. INTRODUCTION

Many patient care goods, such as bed linens, personal protection equipment, dressings, and implantable surgical devices, are made of medical textiles. [1] Because they cover so much of the body, surgical gowns are the most important part of the surgical clothing system and, in addition to protecting the healthcare staff; they also affect their comfort and consequently the operational success. It's composed of cotton muslin fabric, which is a loosely woven, permeable, and reusable cloth. In the nineteenth century, three varieties of fabric were extensively employed. [2] Surgical clothing is a type 2 medical textile used to prevent bacteria from being transferred from surgical staff's skin to the air in the operating room and to protect surgical workers from bodily fluid contact. [3] Microorganisms cannot penetrate the body of a healthy person if the cloth used to make the surgical gown has pore diameters smaller than the germs' size. [4] A surgical gown can be made up of numerous layers, either as a single product or as a collection of products. Microorganisms will have a harder time penetrating multiple layers of the same fabric than one. [5] Along with offering comfort, it's also critical to ensure complete virus protection. The development, application, and quality evaluation of surgical gowns, as well as their requirements, are discussed in this article. It also discusses the various types of gowns available.

2. PERFORMANCE CRITERIA FOR SURGICAL GOWNS

Doctors and nurses use surgical gowns in the operating room to prevent the transfer of germs and bodily fluids from the surgical staff to the patient, as well as from the patient to the surgical staff. [2, 6] The surgical team must be protected; hence these gowns must have certain protective properties. The majority of the performance requirements for surgical gowns are spelled out in detail in the guidelines. The following are the performance characteristics that these aprons must have:

Surgical gowns must be resistant to diseases and infections while still allowing for adequate mobility. They should provide patients adequate mobility while avoiding rubbing, chafing, ripping, and linting. They should fit snugly but not tightly. Because there is generally surplus fabric, the gowns must endure continuous tugs on the fabric during ordinary motions. Because hospitals will only stock a limited number of sizes, with a restricted range of sizes, these gowns should be designed to fit a variety of body shapes and sizes. The gowns must help maintain the sterile zone required for patient safety by controlling bacteria released into the theatre. They must allow for easy donning and doffing without contamination while avoiding any openings that could allow the barrier to be breached. Gowns must be durable enough to survive the intended use life of the garment, whether single or multi-use. While ventilating the surgeon's elevated body temperature, they should resist fluids and must be done at a reasonable cost. [6]

3. SURGICAL GOWN MATERIALS

3.1 Cotton/Cotton–Polyester

Surgical gowns are traditionally made of these fabrics. These are comfortable to wear, but fluid or viruses can flow through since the pore size is large enough, resulting in no barrier effect unless special precautions are taken, such as repeated impregnation with hydrophobic agents to maintain a uniform barrier; even then, there is no guarantee that it will be virus-proof while maintaining the desired level of comfort. Textiles comprised of a combination of polyester and long-staple cotton that have been treated with a liquid-repellent fluorocarbon are also available. Because the repellent finish wears off after each wash, it must be reapplied on a regular basis to maintain the barrier effect. [5]

3.2 Polyester

Despite the fact that surgical gowns can be made from a variety of fibers, polyester appears to be the only one that meets the competing demands of comfort and barrier. Polyester surgical gowns are used for short surgical procedures involving a modest amount of liquid. [4]

3.3 Polyethylene

Polyethylene is a versatile material that is utilised as a fibre in the manufacture of medical garments. Polyethylene is a thermoplastic that is lightweight and has a variable crystalline structure. And it's made by polymerizing the monomer ethylene (or ethene). Low-Density Polyethylene (LDPE) is a very flexible polymer with exceptional flow characteristics, making it perfect for gowns and other plastic film applications. When stretched, LDPE has a high ductility but a low tensile strength, as indicated by its inclination to stretch. Polyethylene, like every other polymer, has benefits and drawbacks that make it more appropriate for certain applications than others. [7]

4. DESIGN ANALYSIS OF SURGICAL GOWNS

Different gowns are now made to handle various procedures. Aprons come in a variety of styles and prices, depending on the level of protection they provide. A systematic, structured approach to design and development is required for the effective creation of practical garment goods. Before arriving at a design solution, an effective, integrated method begins by investigating the design problem, developing design requirements, and critically analyzing those needs. [2,5]

4.1 Sleeves and Cuffs

Raglan sleeves and set-in sleeves are the two most common types of sleeves found on gowns now on the market. The most frequent form of sleeve used in surgical gowns is the raglan sleeve, which is distinguished by a diagonal seam line running from the neckline to the armpit. A set-in sleeve is another sort of sleeve structure that can be found in various garments. The set-in sleeve is more complex to make and restricts the wearer's range of motion. The set-in sleeve is too tight, causing excessive garment pressure under the armpit. [2,5]

4.2 Neck Closures

The gowns' necklines are secured with a variety of methods, including ties, snaps, and hook and loop neck closures, all of which allow for some fit adjustment. The most common method is to use a tie back closure. On the back, a tie band extends along the border of the neckline and inside the neckline towards the shoulder. The comparable ties are located outside the left back piece, at the shoulder or at the edge. Reusable surgical gowns made of slippery textiles tend to slide down while used. Over time, the back tie(s) may loosen, enabling the surgical gown to slide down and causing pain to the medical personnel. [2,5]

4.3 Sizing/Fit

One of the most important factors for surgeon protection and comfort is sizing/fit. Gowns should enable the operator to move easily, fit a variety of body shapes and sizes, and be easy to put on and remove without contaminating the worker or the environment. If gowns aren't correctly fitting, blood can easily get on the skin or other garments. In addition to the universal fit, surgical gowns are now available in a number of sizes (medium, large, extra-large, etc).One-size-fits-all gowns are designed to accommodate a wide range of body types and sizes. [2,5]

5. TYPE OF SURGICAL GOWNS

5.1 Traditional Reusable Materials

All-cotton muslin (140-thread-count muslin) is a loosely woven fabric that is soft, absorbent, drape able, and extremely porous. There is no liquid resistance in this substance because it is easily porous. It also has a proclivity for abrading and generating lint. Blended sheeting is made out of a polyester and cotton blend that has permanent press properties yet performs similarly to muslin in terms of performance. Finally, there's T280 barrier, which is a tightly bound cotton or polyester-and-cotton mix fabric with a water-repellent chemical treatment (175 to 280 thread count), was the first reusable fabric with a water repellent chemical finish. With repeated wash cycles, however, the barrier to liquid penetration decreases. [8]

5.2 Single-Use Materials

Nonwoven fabrics, either alone or in conjunction with materials that provide extra liquid penetration protection, are most typically used in single-use surgical gowns and drapes (e.g., plastic films). Nonwoven fabrics are made up of a variety of natural (e.g., wood pulp, cotton) and synthetic (e.g., polyester, polyolefin) fibers by the use of specialized fiber types, bonding processes, and fabric treatments, that may be created to attain specified attributes. Spun lace, a hydro entangled material made up of wood pulp and polyester fibers; spun bond/melt blown/spun bond, a fabric made up of three thermally or adhesively bonded layers (spun bond provides strength, melt blown provides a barrier); and a nonwoven fabric consisting of wood pulp or a combination of polyester and wood pulp fibres that is wet-laid. [8]

5.3 Reinforcement of Reusable and Single-Use Product

Reinforcement is often used to strengthen or improve the qualities and performance of both reusable and single-use items. For some surgical gowns and drapes, the barrier properties of a single ply (i.e., single sheet or layer of fabric) of material may not be adequate; in these cases, additional materials are sometimes added in the form of extra layers of material, coatings, reinforcements, or laminates. Reinforcement can also be used to increase absorbency, non-slip capabilities, and other desired characteristics. 12 Reusable and single-use objects employ standard (one layer) fabric, reinforced fabric (second layer of fabric used to reinforce base material), zone impervious fabric, and impermeable fabric. [8]

6. CONCLUSION

To create and maintain a sterile environment, surgical gowns are used. The surgical gown's primary function is to keep fluid out. Fluid-repellent chemicals are applied to cotton muslin and tight weaves have traditionally been used to make woven reusable surgical gowns. The majority of gowns are now made of polyester or a polyester-cotton blend. Structural analysis is conducted to characterize the market and assess current product design elements. Product design difficulties relating to comfort are especially concerning for protective clothing, as comfort is a fundamental product need. When the risk of exposure to bodily fluids is low, other gowns may be used. Surgical personnel should decide which type of gown to wear in various operating room scenarios. Any instances of penetration would indicate the need for a higher level of security.

REFERENCES

[1] Vozzola, E., Overcash, M. and Griffing, E. (2020). "An Environmental Analysis of Reusable and Disposable Surgical Gowns". *AORN Journal*, 111(3), 315–325.

[2] Atay, Y., Pamuk, O., Boyaci, B., Yildiz, E.Z., Göksel, T., Metin, D.Y. and Topbaş, O.A. (2021). "New Approach to Surgical Gowns". In A. Agrawal and S. Kosgi (Eds.), Healthcare Access. Intech Open. https://doi.org/10.5772/intechopen.98563

[3] Rogina-Car, B. (2018). "Functional Design and Construction of Reusable Surgical Gowns Considering Microbial Barrier", *Textile & Leather Review*, 1(2), 57–62.

[4] Behera, B.K. and Arora, H. (2009). "Surgical gown: a critical review", *Journal of industrial textiles*, 38(3), 205–231.

[5] Gokarneshan, N., Rachel, D.A., Rajendran, V., Lavanya, B. and Ghoshal, A. (2015). "Surgical gowns—techno economic aspects and innovations", *In Emerging Research Trends in Medical Textiles Springer*, Singapore, 97–119.

[6] May-Plumlee, T. and Pittman, A. (2002). "Surgical gown requirements capture: a designanalysis case study". *Journal of Textile and Apparel Technology and Management*, 2(2), 1–10.

[7] Medical Gowns Materials, https://munglobal.com.au/resources/knowledge-base/personal-protective-equipment/medical-gown-materials

[8] Rutala, W.A. and Weber, D. J. A. (2001). "Review of Single-Use and Reusable Gowns and Drapes in Health Care". *Infection Control & Hospital Epidemiology*, 22(4), 248–257.

Nanotechnology Application in Antimicrobial Textiles and Its Significance in Post-pandemic Scenario

Chikoti Vakya Priya, Sahithi Natva and K. Ram Mohan*
National Institute of Technology, Hyderabad, Telangana, India
E-mail: *kram.mohan@nift.ac.in

ABSTRACT: *The antimicrobial textile is largely contributory in today's world as they inhibit or kill the growth of microbes in the garment. It can be widely used in medical garments to prevent spread of infections and also in regular garments to certain extent. Use of these fabrics is even more pertinent in the present pandemic period when spread of COVID virus in its crowning, as this fabric helps in preventing or minimizing the spread. The frontline workers are most frequently prone to pathogen attacks; hence garments made using this type of textile are most useful for their protection. These textiles are very beneficial in the current circumstances where safeguarding from the toxic microbes is imperative for the frontline workers. This paper provides the reader with a detailed understanding of the role of nanotechnology and its standing in the present magnitude of COVID Virus, the types of applications and incorporation of nanotechnology in the antimicrobial textiles along with the treatments, chemicals, and substrates used. The procedure of manufacture and their availability in India are also covered in the study.*

Keywords: Nanotechnology, Antimicrobial, Textiles, Post-pandemic.

1. INTRODUCTION

Microorganisms thrive in textile fabrics, which provide an ideal environment for their growth and proliferation. Natural fibers include fundamental chemical elements that supply nourishment to microorganisms and hence encourage their development. The proliferation of microorganisms in the textile materials causes a slew of issues, including unpleasant odors, fabric deterioration, and stains, which leads to affect the wearer's health. As a result, it's critical to provide textile materials with an antibacterial function in order to safeguard the wearer's health. In textile applications, a variety of antimicrobial agents have been employed such as silver nanoparticles. One of the finest antibacterial properties of these silver nanoparticles is that they cover a wide spectrum of harmful microorganisms. They react with the sulfur-based proteins in bacteria's cell walls, inhibiting their metabolism and killing them.

Silver nanoparticles have been efficiently used to treat cotton, wool, and silk materials to achieve the necessary antibacterial effects. Metal oxides such as ZnO, MgO, CaO and TiO_2 are of special importance among inorganic materials since they are not only stable under rigorous process conditions but are also usually recognized as safe material for humans and animals. Because of their antibacterial qualities, zinc and silver oxide nanoparticles have been proposed as a feasible approach for preventing catching infections.

From the advent of Coronavirus, the number of antibiotic-resistant microbes is escalating, as is the prevalence of diseases caused by these germs. Figure 1 is an illustration of how antimicrobial fabric deals with bacteria, fungal & virus. The whole world population has fallen prey to this microbe generated spread and this wave can only be inhibited by properly handling their proliferation.

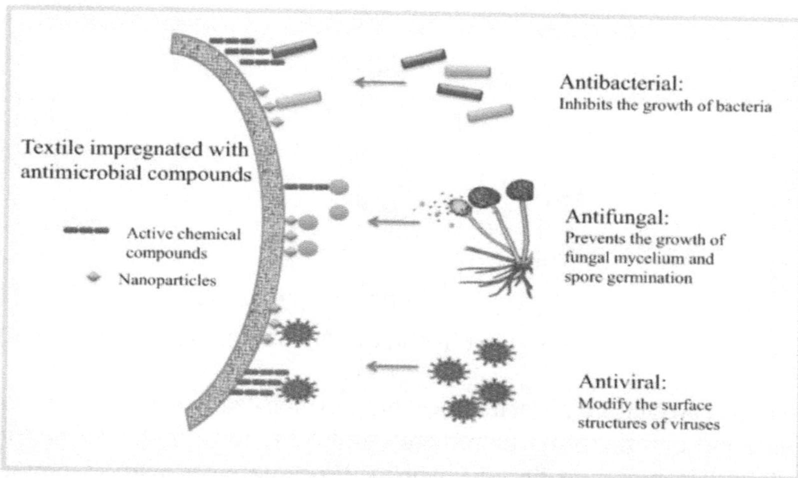

Fig. 1: Antimicrobial Fabric Dealing with Bacteria, Fungal & Virus

¤ *Image source:* (https://link.springer.com/article/10.1007/s00289-021-03826-3#Tab)

This pandemic scenario has driven people keener towards antimicrobial fabrics as they serve as a massive weapon in fighting diseases and stay safe while performing their respective duties. It has been aimed to summarize the technology, process of application, use of nanotechnology in antimicrobial fabric and their availability in India. [1-19]

2. WORKING PRINCIPLE OF NANO FINISH

Antimicrobials are substances that inhibit or destroy microbes. It also improves microbial metabolism, reducing smell, stains, and color loss while increasing fabric durability.

Antimicrobial finishes are divided into two categories based on mechanism of activity:

1. *Biostatic finish*—this antimicrobial finish slows or stops bacteria from multiplying without harming them. The fabrics are biostatic fabrics.
2. *Biocidal finish*—this antimicrobial finish destroys or kills germs, and the fabrics are biocidal fabrics.

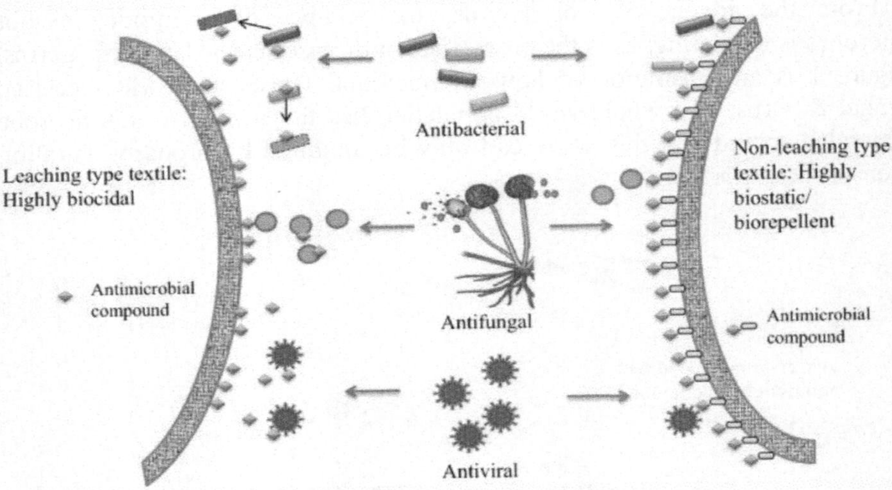

Fig. 2: Types of Antimicrobial Textile

Image source: (https://link.springer.com/article/10.1007/s00289-021-03826-3#Tab)

The antimicrobial fabric can be leaching or non-leaching depending on the treatment and antimicrobial ingredient employed, and it can also be biocide or biostatic depending on the mechanism. Biostatic fabrics are recommended for clothing because they retain the skin's normal bacterial flora and have no negative impact on human skin, whereas biocidal fabrics are favored for medical and environmental purposes.

Most of antimicrobial compounds used in the textile sector works on controlled release or leaching process in the presence of moisture. Additives can be injected into the fibre matrix during spinning extrusion, or they can be applied to the textile surface through a traditional coating or impregnation method like padding or exhaustion. [1-19]

3. DIFFERENT ANTIMICROBIAL FINISHING AGENT

Some of the types of finishes are mentioned below.

3.1 Finish with Zinc Oxide

The textiles soaked in solution of 2-propanol dispersion zinc oxide nano-particles for 10 minutes. After that the textiles were compressed on padded mangle and oven-dried for 15 minutes at 130°C at air pressure. After that, the treated materials are washed five times, laundry was done at 30°C using a reference detergent that did not include optical brighteners and the washing cycles were completed.

3.2 Finish with Titanium Oxide

Pad dry curing method is normally used for application of TiO_2 nanoparticle in cotton and polyester fabrics. The material liquor of ratio 1:20 is applied on the facing side with a 1% acrylic binder. The extra solution is removed and then it is cured and dried for about 3 minutes at 150 degrees. Antibacterial activity of titanium dioxide nanoparticles treated fabrics has been summarized Table 1.

Table 1: Antibacterial Activity of Titanium Dioxide Nanoparticles Treated Fabrics

Fabric samples	*Staphylococcus aureus*	*Klebsiella pneunomiae*
Untreated	No reduction	No reduction
Woven 100% cotton	85%	64%
Woven 45/55 polyester/cotton	93%	78%
Knitted 100% cotton	75%	58%
Knitted 45/55% polyester cotton	79%	74%

Source: https://www.hindawi.com/journals/isrn/2012/193836

3.3 Finish with Silver Oxide

The powdered silver nanoparticles are dissolved in water to obtain a silver level of 50 parts per million (ppm). The procedure of exhaustion is used to apply this to cotton and wool fibers. The materials are then treated with boiling water for 45 minutes before washing with regular water. The treated textiles durability is tested for 5, 10, and 20 washes using the AATCC 124–2001 method.

3.4 Finish with Silver Nano Colloids

The textiles were soaked for 10 min in the solution before padding using a twofold dip and nip process with a 70% expression. After that, the padded textiles are air dried and cured for 20 minutes.

3.5 Finish with Polyethylene Terephthalate

PET textiles are treated to single- and double-sided corona air treatment at atmospheric pressure using specially designed equipment. Fabrics are put on a backing electrode roller (2 mm apart) with a silicon covering that rotates at a 4 m/min speed. The bacterial reduction of blank-colored textiles with nano silver particles (80 nm) was investigated using the ASTM-E 2149-01 technique. Corona plasma treatment was utilized to prepare the surfaces of PET textiles in order to improve their adhesion to nano silver particles. [1-19]

4. USE OF ANTIMICROBIAL FINISH IN DIFFERENT FIELD

Between 2020 and 2025, the antimicrobial textile market is estimated to increase to a CAGR of over 4%. In 2020, India's garment market will be valued $53.7 billion, making it the world's sixth biggest. As a result, it only makes sense for manufacturers and brands to incorporate new age technologies such as "germ protection", which will help improve sales and reinvigorate customer interest.

Siyaram's, a men's fashion textile business, and Health Guard, a global leader in non-invasive healthcare, have released an anti-corona fabric collection that has been tested by WHO (World Health Organization) authorized labs to combat the pandemic's spread. HeiQ Viroblock NPJ03 was launched by D'Decor Home Fabrics and HeiQ, a Swiss textile innovation firm. It's an antiviral and antibacterial textile treatment that decreases viral and bacterial infectivity on treated surfaces using a unique mix of silver and vesicle technologies.

Manufacturers of Antimicrobial textiles: Donear Industries, Welspun India, Arvind and Vardhman Textiles are just a few of the firms that make similar products. Antimicrobial textiles are also produced by Aditya Birla Yarn and can be utilized in end user garments.

Major evolutions are taking place in this field, the greatly renowned 'Surti Sari' has now become infection free. Laxmipati Industries has started coating sarees, kurtis and ethnic wear with the disease-killing chemicals to make them germ resistant. Arvind Ltd has launched the antiviral technology infused clothing. Chiripal Group has introduced the new terry towel and denim range that is antibacterial and antiviral in nature. More and more textile leading houses are all venturing into this practice and this is a kind of fabric that sees a lot of scope of improvement, its use in some of the different industries are discussed below. [1-19]

4.1 Antimicrobial Finish in Medicine Field

Every year, millions of people around the globe contract a healthcare-acquired infection (HAI), making HAIs a global problem that contributes significantly to mortality, morbidity, and hospitalization costs, especially since many pathogens have developed high resistance to the current antibiotic arsenal.

The goal is to protect health professionals from coming into direct touch with a patient's secretions and as a result microbes are transfers among patients, doctors and other health workers. Usage of antimicrobial treated uniforms, bed sheets and covers creates microenvironment, where microbes are being destroyed.

4.2 Antimicrobial Finish in Dermatological Purposes

Antimicrobial fabrics are used in a range of settings, including first aid, therapeutic procedures, and hygiene standard. Antimicrobial textiles could also be used as bio-functional textiles that combine sophisticated pharmaceutical nanocarriers with standard fabrics to develop wearable medicine delivery systems. Silver's ability to adhere to cotton cloth and penetrate it suggested that they may be employed as a wound treatment. Ketoconazole and cyclodextrin were included in this cotton wound dressing, demonstrating a regulated and delayed release of these antifungal chemicals to kill skin fungus.

4.3 Antimicrobial Protection for Automotive Industry

The drivers touch all the germy surfaces and touch the interiors. Because car interiors are not cleaned as regularly as other surfaces and parts of our daily lives, this problem is exacerbated in car sharing, rental vehicles and commercial vehicles with several drivers. Hence antimicrobial textiles are instrumental in this sector.

4.4 Antimicrobial Protection for Military and Defense

For the future force warrior, researchers devised a textile finishing technique to manufacture sophisticated bio-textiles that will include both broad-spectrum "leaching" and "binding" antimicrobial chemicals as skin/wound infection barriers. The project resulted in a high-kill rate, long-lasting organic antibacterial chemical that was applied to troops clothing, including Battle Dress Uniforms (BDUs), t-shirts, and socks.

4.5 Antimicrobial for Active-wear

Excessive sweating occurs as a result of this exercise, which promotes bacteria development and increases the presence of body odor. Natural oils from our skin might enter into the cloth when we exercise and sweat. These oils develop bacteria that produce the stench we detect. Antimicrobial chemistries are used by well-known companies like Nike, Reebok, and Under Armour in many of their goods to keep sportswear fresh and odor-free for longer.

4.6 Antimicrobial for Food Packaging

Antimicrobials may be included into packaging materials in a variety of ways, including direct integration of antimicrobial agents into polymers or the application of antimicrobial coatings to polymer surfaces. Antimicrobial packaging permits the integrated agent to release in a regulated or sustained way, inhibiting bacteria growth throughout transit and storage.

4.7 Antimicrobial for Intimate Apparel

Bras and related garments were in high demand by the woman's apparel business as breathable, antibacterial, and antimicrobial breast support; a breast support system was developed. Apart from that, men's designer antimicrobial underwear is also in use. Four 140D antimicrobial nylon strands were used in a weaving procedure in a patent for men's underwear.

5. PANDEMIC SCENARIO

The epidemic has amplified the effects of these fabrics. Henceforth, such expansion will continue long after one of the most effects of the COVID-19 epidemic have abated. As a direct consequence of the COVID-19 pandemic's heightened concern about hygiene and welfare, consumer awareness of the benefits and availability of antimicrobial textiles has developed. As a result, current estimates indicate that the market for antimicrobial textiles will grow at a rate of around 10% per year in the medium term, compared to expansion of around 7% per year prior to COVID-19's breakdown.

Antimicrobial face masks, antimicrobial at hleisure clothing, outdoor gear and sportswear, and antimicrobial uniforms and work wear are among the items in greater supply. Interestingly, antimicrobial materials are increasingly being used in a diverse range of garments in response to client's needs about hygiene and well-being.

Antimicrobial textiles, in fact, could be utilized in nearly any agent application, thanks to the ease with which antimicrobial treatments may be applied to a wide range of materials. Consumers will continue to prefer functional fabrics in general, and will be willing to pay more for clothing that integrates various functionalities such as water repellency, breathability and microbial resistance. Antimicrobial fabrics, on the other hand, will confront a variety of obstacles, including those related to environmental sustainability. [1-19]

6. CONCLUSION

Antimicrobial textiles have been developed using both synthetic and natural fibers. Chemicals, metal-based nanoparticles, plant and animal-derived substances, dyes and mordants can all be used to impart antimicrobial characteristics. Antibacterial effects against bacteria, fungus and viruses may be limited or wide in coated textiles. The amount of antimicrobial ingredients to be coated may be determined using MIC values, which impacts the efficiency and cost of prepared textiles. As a result, future study should focus on the potential of natural antibacterial agents.

With the increased usage of antimicrobial textiles in the home, a well-planned and regulated system for antimicrobial textile handling and disposal is needed. A consistent and efficient approach is essential to avoid this becoming an issue similar to plastic waste management.

REFERENCES

[1] 5 Benefits of Antimicrobial Fabric, MICROBAN, https://www.microban.com/blog/5-benefits-of-antimicrobial-fabric

[2] Antimicrobial protection for automotive industry, BioCote, https://www.biocote.com/environments/transport/antimicrobial-protection-for-automotive-industry

[3] Borkow, G. (2019). "Antimicrobial medical textiles used in clinical settings may play an important role in reducing healthcare-acquired infections". https://www.researchgate.net/publication/338946653_Antimicrobial_Medical_Textiles_Used_in_Clinical_Settings_May_Play_an_Important_Role_in_Reducing_Healthcare-Acquired_Infections

[4] Daniela, P.Z., Rozman, U., Pal, E., Gonc, V. and Sonja, T.S. (2017). *"Efficiency of medical workers' uniforms with antimicrobial activity"*, https://www.intechopen.com/chapters/55317

[5] Gavrin, A.J., Gonyer, R.G., Blizard, K.G. and Santos, L. (2004). *"Medical textiles for uniform bacteria mitigation"*. https://www.researchgate.net/publication/235151988_Medical_Textiles_For_Uniform_Bacteria_Mitigation

[6] Gokarneshan, N. (2012). *"Influence of nanofinishes on the antimicrobial properties of fabrics"*, https://www.hindawi.com/journals/isrn/2012/193836/

[7] Jim, K. (2017). "Nano-technology prescription for antimicrobial protection". https://www.textileworld.com/textile-world/features/2017/09/nanotechnology-prescription-for-antimicrobial-protection

[8] Antimicrobial Active wear—What is the real Purpose? NICCA, https://www.niccausa.com/antimicrobial-activewear-real-purpose/

[9] Perkas, N., Perelshtein, I. and Gedanken, A. (2020). "Coating Textiles with Antibacterial nanoparticles using the sonochemical technique", *Nanotechnology in Textiles*, Taylor & Francis Group. https://www.taylorfrancis.com/chapters/edit/10.1201/9781003055815-26/coating-textiles-antibacterial-nanoparticles-using-sonochemical-technique-nina-perkas-perelshtein-aharon-gedanken

[10] Niyati, P. and Melvyn, T. (2021). "Gujarat: Antiviral fabrics catch textile companies fancy", https://timesofindia.indiatimes.com/city/ahmedabad/gujarat-antiviral-fabrics-catch-textile-companies-fancy/articleshow/76488587.cms

[11] Beyth, N., Houri-Haddad, Y., Domb, A., Khan, W. and Hazan, R. (2015). "Alternative antimicrobial approach: nano-antimicrobial materials". *Evidence-Based Complementary and Alternative Medicine*, 2015, Article ID 246012, 16 pages, http://dx.doi.org/10.1155/2015/246012

[12] Pal, S., Mondal, S., Pal, P., Das, A., Mondal, D., Chaudhuri, A., Panda, B. and Maity, J. (2020). "Applications of nanotechnology for antibacterial finishing textiles: a review". *Sensor Letters*, 18, 1-12, DOI:10.1166/sl.2020.4260

[13] Gulati, R., Sharma, S. and Sharma, R.K. (2021). "Antimicrobial textile: recent developments and functional perspective". *Polymer Bulletin,* 2021, https://doi.org/10.1007/s00289-021-03826-3

[14] Activewear, Sciessent, https://www.sciessent.com/markets/textiles-and-apparel/active-wear/

[15] Sristi, S. (2021). "Antimicrobial finish in textile during COVID-19", https://www.fibre2fashion.com/industry-article/9205/antimicrobial-finish-in-textile-during-covid-19

[16] Apparel manufacturers evaluate antimicrobial clothes to get back to business, *Techtextil, 2020*, https://techtextil-india.in.messefrankfurt.com/mumbai/en/facts-figures/Techtextil_India_Blog/techtextilindia-introductoryblog/2020/Apparelmanufacturersevaluateantimicrobialclothes.html

[17] Antimicrobial textiles: Promoting hygiene during the Covid-19 Pandemic and beyond. *Textiles Intelligence Ltd.*, 2021, https://www.researchandmarkets.com/reports/5331889/antimicrobial-textiles-promoting-hygiene-during

[18] Vhanbatte, S.B., Landage, S.M., Wasif, A.I., Dansena, B. and Karche, N.V. (2017). "Nanotechnology for antimicrobial finishing of textiles", *International Journal of Advanced Research in Engineering and Applied Sciences*, 2017, 6(6), 14–23.

[19] Zille, A., Almeida, L., Amorim, T., Carneiro, N., Fátima E.M., Silva, C.J. and Souto, A. P. (2014). "Application of nanotechnology in antimicrobial finishing of biomedical textiles". *Materials Research Express*, 1(3):032003, DOI:10.1088/2053-1591/1/3/032003

Opportunities and Challenges for New Entrepreneurs in Medical Textiles

Ketankumar Vijaykumar Vadodaria*
National Institute of Design, Textile Design, Paldi Campus, Ahmedabad, Gujarat, India
E-mail:* kvarticles@live.co.uk

ABSTRACT: *The aim of this study is to get an overview on medical textiles in India. The Indian textile industry can manufacture a diverse range of products from fibre, yarn, fabric, and apparel manufacturing to the latest technical textile manufacturing. Almost 16 different domains of technical textiles are identified, and new segments are added with the new inventions in advanced textiles and consumer needs. Though India has been already manufacturing technical textiles in certain technical textiles domains, it still lacked the required capacity and advancement. Still, many technical textile products are imported. Recent initiatives and policies are focused on boosting the technical textile industry in India. One of the emerging and fastest-growing segments in the field of technical textiles is medical textiles. With the outbreak of COVID, medical textiles had become a national and international emergency. Medical textiles is one of the sunrise technical textile domains where India is still an infant. Medical textiles products range spread from simple product like wipes to complex products like vascular grafts. The demand for medical textile products are increasing every year as average lifespan is increasing. There are huge scope to produce medical textile products in India for both domestic and the international market.*

Keywords: Technical Textiles, Medical Textiles, Challenges, Opportunities, Market.

1. INTRODUCTION

Medical textiles are the category of technical textiles which are engineered and designed for clinical, therapeutic, medical, surgical, health and hygiene purposes. Technical textiles are the structures made by combining different raw materials and textile structures to meet the specific purpose. Technical textiles are classified based on specific domain of end use such as— Medical textiles, Agro Textiles, Build Textiles, Cloth Textiles, Geo Textiles, Home Textiles, Industrial Textiles, Mobil Textiles, Oeko Textiles, Pack

Textiles, Protective Textiles, Sport Textiles. More new classes of technical textiles are being identified and added with new innovations such as space textiles, etc. Medical textiles products range spread from simple product like wipes to complex products like vascular grafts. The demand for medical textile products are increasing every year as average lifespan is increasing. [1-12]

2. CLASSIFICATION OF MEDICAL TEXTILE PRODUCTS

2.1 Medical Textiles can be Divided into (based on the market)

- *Consumer products:* Sanitary napkins, baby diapers, wipes, adult diapers.
- *Hospital products:* Surgical gowns, surgical drapes.

2.2 Medical Textiles Classification (based on the end use)

- *Implantable:* Inside the patient's body (Vascular Graft, Meshes, Sutures, Abdominal Walls, Dura Fabric/patches, Tendons/Ligaments, etc.)
- *Non-Implantable:* On the patient's body (Bandages, Gauzes, Compression garments, Waddings, Wound dressings, etc.)
- *Extracorporeal Device:* Besides the patient's body (Artificial kidneys, Artificial heart and lung, etc.)
- *Healthcare textiles:* For the body (Usually hospital products)—(PPE kits, Medical wipes, Maternity pads etc.)
- *Hygiene textiles:* For the body (Usually consumer products)—(Sanitary Pads, Home wipes, Diapers, Mops, etc.)

2.3 Medical Textiles Classification (based on the materials)

- *Raw Material:* Cotton (Cotton wool roll), Polypropylene (PPE kits), Nylon/Silk (Sutures).

2.4 Medical Textiles Classification (based on textile structure)

- *Fibre:* Cotton-Wool roll, Earbuds, Pleats, wadding, fillings.
- *Yarn/Filament:* Sutures, Dental floss, Dialysis membrane filament.
- Fabric (Woven, Braided, Non-woven, Knitted, Lace, Embroidered). Table cover, Bandages, Meshes and other implantables, Sutures, Ligaments, Grafts.
- *Stitched Garment:* PPE kits, Dead body covers, Isolation Tents, Bed sheets/pillow covers.
- *Products:* Diapers, Sanitary napkins.

2.5 Medical Textiles Classification (based on Finish/treatment)

- *Finishes:* Antimicrobial, Anti static, Sterile. [1-12]

3. PROPERTIES OF MEDICAL TEXTILES

The desired properties of medical textiles can depend on the enduse.

- Barrier performance/Breathability—Breathable but barrier to air borne and water borne particles/microbes (PPE kits)
- Hydrophilic (Wound dressings)/Hydrophobic (PPE kits)
- There are antimicrobial products, which are bacteria, viral, fungal, mould resistance etc (Wound dressings, Wipes)
- Biocompatibility (Implantables)
- Cell adhesion (Tissue engineering—Wound dressings)/Cell non-adhesion surface (Vascular graft inner surface, Meshes)
- 2D & 3D structures (Bandages, Meshes)
- Sterilizability (Fabric structure and properties should not get altered while steralization)
- There are biodegradable (Short terms Sutures, Implants)/Non bio-degradable (Long terms Sutures, Implants)
- Elasticity/non elasticity
- Strength. [1-12]

4. OVERALL OPPORTUNITIES IN MEDICAL TEXTILES

As the demand for medical textile is increasing, the opportunities in every field including raw material manufacturing, laboratories to test these materials, manufacturing, etc is also increasing.

- Though there is no exact figure of the amount of medical textile product imported in India but it may be around 80%. So there is huge opportunity for the import substitutions.
- As mentioned earlier that the population is very high in India but due to lack of awareness the consumption of medical textile product is low. So, there is opportunity increasing the consumption of disposable products also
- Higher child birthrates—the products related mother and child care.
- Modernization in hospitals and new hospitals are coming, hence there is chances for increasing the consumption of medical textiles.
- It is recommended by insurance companies to maintain a proper environment at hospitals to reduce risk of HAI.
- Awareness is increasing in India to use of wipes, sanitary napkins, diapers for cleanliness & hygiene environment.

- There is huge opportunity in India for manufacturing as availability of basic raw materials is easy.
- Different initiatives taken by Indian govt. as MoD, MoHFW, MoLE, MoR, MoHA will help start-ups. For example, Centre of excellence at SITRA for medical textiles was established and working for Research and development, formation of testing standard, testing/characterisation of products, etc. The other initiative by Govt. of India like subsidies, special zones (Meditech zone, Visakhapatnam), ATUF, TMTT, etc., could be availed by the start-ups.
- Many NGOs as well as government agencies are working for creating awareness.
- Easy availability of manpower and electricity could also help in manufacturing cheaper price.
- Mandating purchase from MSMEs.
- Schemes such as Swachh Bharat Abhiyan, Ayushman Bharat. [1-12]

5. SECTOR WISE OPPORTUNITIES IN MEDICAL TEXTILES

5.1 Implantable

India is almost dependant on imports for costly life saving products, which may not affordable for an average Indian. The cost of many of those products will come down if those products are manufactured by Indian companies.

5.2 Non-implantable

Non-implantable products are those, which are used on the skin. Though there is less profitability but there is high demand.

5.3 Extracorporeal Devices

These are Implantable products, which are costly and life saving. We are highly depended on import for these products.

5.4 Hygiene Textiles

Demand is high and increasing day by day, investment required is low.

5.5 Healthcare Textiles

Demand is high and investment required is low. [1-12]

6. CHALLENGES FOR MEDICAL TEXTILE BUSINESS

- Licensing problem for production and selling Medical device products in India is a constrain (New medical device rule is implemented recently).
- Availability of standards for testing the products is less in India. SITRA has developed standards for many medical textile products.
- Raw materials (high performances fibre) Kevlar, Nylon66, Aramid, etc., are not manufacture in India.
- Scattered industry.
- Lack of trained skilled manpower with interdisciplinary knowledge.
- Disposal of medical textile wastes. [1-12]

7. INITIATIVES REQUIRED

Government can take initiative for both Rolled goods and converting manufacturing. Which may contain research, pilot plant production; ready good products for global markets can be developed.

1. Development of new Infrastructure; more specialized zones like Medtech Zone in Visakhapatnam.
2. Focus on Indian IPR regime,

   Highly equipped with modern facilities patent offices and online patent filing facilities is in demand.
3. More benefits in tax incentives, subsidy by government at every stages like production, machine imports, machine spares and machine manufacturing is in demand.
4. More and more medical textile manufacturing clusters is required to produce, test and market their products.

   Encouragement in terms Subsidy and easy paybacks for the start-ups should be provided to develop more indigenous machines. [1-12]

8. STATISTICS ON MEDICAL TEXTILES IN INDIA

The total market size of medical textiles in 2017 was $780 million with a growth of 8–9% CAGR from $600 million in 2014. The market share of only surgical dressings and surgical sutures was more than 50 per, which was $380 in 2014. The market size of different medical textiles, has been shown in Table 1. [1-12]

Table 1: Market Size of Different Medical Textiles [2]

Medical Textiles	Market size	Annual growth rate (%)
Surgical dressings	$154.43 million	5-10
Healthcare textiles	$26.72 million	13-16
Sutures	$98 million	15-20
Sanitary napkin	-	8-10
Diapers	$17.4 million	5-10
Medical implants and devices	$154.43 million	10-15

9. CONCLUSION

From the above discussion it is very prominent that medical textile field is a thriving field. Business opportunities in medical textiles are enormous. Career and Business Opportunities in the field of Medical Textiles is summarized in Figure 1. Now it is right time to utilize this for better and brighter human society. [1-12]

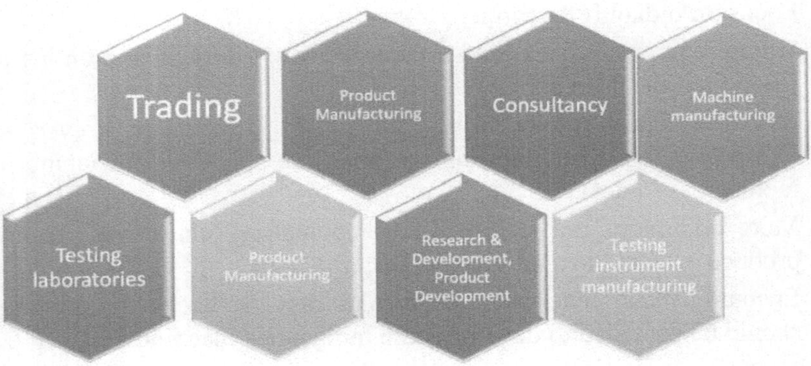

Fig. 1: Career and Business Opportunities in the Field of Medical Textiles

REFERENCES

[1] Office of the textile commissioner, Baseline survey of the Technical Textile industry in India, 2018, https://www.technotex.gov.in/BSTT_Rev_Final_report %20180420 16.pdf

[2] Alagirusamy, R. and Das, A. (2010). "Technical textile yarns-Industrial and medical applications", Woodhead publishing limited, UK.

[3] Medical Textiles set to grow at 9% CAGR in India, 2018, https://www.indiantextilejournal.com/News.aspx?nId=oBaxRIEDzihNK+u7Yjfd/g==&NewsType=Medical-textiles-set-to-grow-at-9%-CAGR-in-India-India-Sector

[4] Global Medical Textiles Market Expects Substantial Growth (2018). https://www.bizvibe.com/blog/global-medical-textiles-market-expects-substantial-growth

[5] Understanding the Difference between Class I, Class II and Class III Medical Devices, 2018, https://www.innovatum.com/2014/12/understanding-difference-class-class-ii-medical-devices

[6] Apparel, Consumer Markets, Statista (2018). https://www.statista.com/outlook/90050000/119/hosiery/india?currency=usd#market-arpu

[7] Kranthi, K.R. (2013). "Impressions from a cross-section", *fibre2fashion.com*, https://www.fibre2fashion.com/interviews/sector-pulse/central-institute-for-cotton-research-cicr/dr-keshav-raj-kranthi/148-1

[8] Turon, P. (2017). "Trends and challenges in textiles associated with medical devices used in surgery". *International Textile conference.*

[9] Bengaluru scientists develop artificial liver tissue, Bangalore Mirror Bureau, 2015, https://timesofindia.indiatimes.com/city/bengaluru/Bengaluru-scientists-develop-artificial-liver-tissue/articleshow/50293458.cms

[10] Mishra, L. (2017). "Nobel Hygiene to boost diaper capacity", *The Hindu.* https://www.thehindu.com/business/nobel-hygiene-to-boost-diaper-capacity/article17709285.ece

[11] Olivo, T. (2018). "Canopus Enters Indian Wet Wipes Market", *Nonwovens Industry since 1970*, 2018, https://www.nonwovens-industry.com/contents/view_online-exclusives/2018-02-20/canopus-enters-indian-wet-wipes-market/13590

[12] Global notice inviting expression of interest from manufacturers/distributors/marketing agencies requiring Oem supply of surgical sutures, HLL Life care Ltd., 2018, https://www.go ogle.com/url?sa=t&rct=j&q=&esrc=s&source=web&cd=5&cad=rja&uact=8&ved=0ahUKEwju6YXkoYLcAhXaT30KHdHUBTcQFghBMAQ&url=http%3A%2F%2Fwww.lifecarehll.com%2Ffile%2Fdownload%2Freference%2F1bd2caf96a17d892c2c7e9959549cfc7i4iDgA&usg=AOvVaw1SLYdUQpPndthOb4GrEhaw

Application of Smart Textile

Trupti Rathod
Veermata Jijabai Technological Institute, Mumbai, India
E-mail: rathodtrupti446@gmail.com

ABSTRACT: *The present study aims to make a look on the comprehensive understanding of smart textile in terms of preparation, properties and usages in various fields. Civilization has brought tremendous changes in the day to day life of human beings. All things are changing day by day through the efforts of scientific activities of researches. Among these textiles are good examples. During the industrial revolution they showed up one step in their evolution in raw material like synthetic and regenerated fibres, machineries, process like plasma technology, etc. Since the last 20 years, smart or stimuli responsive textile material have been introduced and it will boom in the near future. They are everywhere; from geo textiles in the soil to the outer space in the expedition of the u to sportswear applications.*

Keywords: Smart Textile, COVID-19, Geotextiles, Plasma Technology, Synthetic Material, Regenerated Fibres.

1. INTRODUCTION

Smart textiles are value added textile products, which help keep the wearer cool and track physical condition. They are sometimes also called e-textiles. Smart textiles can sense external stimuli, communicate with other devices by sending the signals and protect us by responding to the external stimuli. There are even textiles that light up and change colour. As this description might suggest, smart textiles tend to be used in areas like health, military and defence, and sportswear, especially for extreme sports. [1-5]

2. DEFINING SMART TEXTILES

A smart textile really is: a cloth that has modern computer-based technology woven into it, which can communicate with smart phones. This will ultimately received the signals and respond according to the programming. Smart textiles could be classified into two categories: aesthetic and performance enhancing. Aesthetic smart textiles are those textile materials, which is mainly used for fun like light and sound integrated into fabric. On the other hand, there are performances enhancing textiles, which sense and monitor our body like temperature, heartbeat, blood pressure, adjusting the breathability of the fabric accordingly. This branch of textile is comparatively younger than other sectors of technical textiles. The key factors, however, are responsiveness to external stimuli and the use of thin technological fibers woven into fabric. The possibilities of smart textiles seem to be nearly endless, and that's without even delving into the simply fun realms of art and fashion design. [1-5]

3. MATERIALS USED IN SMART TEXTILES

Smart textiles use materials like advanced fibers and coatings and many of them are conductive, which means they allow for the flow of electrical currents. The fibre or yarns used for smart textiles may be of metallic, metal inserted or common textile fibres treated with conducting polymers. Fabrics can be woven with stainless steel yarn, silver threads or other precious metals.

Other textiles are woven with optical fibers and LEDs, light-emitting diodes that glow when they come into contact with electricity. The yarns could be coated with active components to acts as sensors. Active circuits can be printed with conductive inks onto the fabric that would respond to changing situations. There are even companies experimenting with growing bio-leather. They use microbes to make a translucent material similar to leather. Just think, some day one might wear clothes made from bacteria! [1-5]

4. WORKING OF SMART TEXTILE

Smart textiles made by incorporating smart materials in normal textile structure either as an attachment or the textile structure become a part of the circuit. The functional properties as per requirement like conductive polymers, encapsulated phase change materials, shape memory polymers and other materials like electronic sensors. In this case the incorporated materials interact with the stimuli in their environment and react accordingly as an actuator. The smart textile which is interact with skin are getting attention of clothing towards smartness But skin is ever-present. Skin is a flexible multilayered, multipurpose organ. It works as a barrier, a container and biochemical manufacturer. It also guards the body from infection, harmful radiation and mechanical and electrical forces. It protects our body by water-repellent and inflammatory response, controls body temperature and fluid balance, absorbing vitamins and cleaning cells, etc. Clothes are called second skin of human beings. Clothing is personal, comfortable, close to the body, especially to the skin and used almost anywhere and anytime. Many clothing manufacturers claim that their product breathes, sense and lives, like second skin, even sense the location as well as pressure from human touch also. Furthermore, smart textiles could be programmed to monitor things like biometrics (measurements of physical attributes or behaviour such as heart rate), which could help athletes, people on diets, and physicians observing patients' vitals. [1-5]

5. RELATION AND DIFFERENCE WITH TECHNICAL TEXTILES

Textile is a material very close to the human being and considered as the second skin of humans, which relation has a relationship. In the current and also in the past 200 years it has shown dynamic changes. From the way we can see the research and innovation on the sector will continue in the coming years. The latest approach is amazingly fuelling the development of new to integrate textiles with other electronic and almost everyone from young to old, male to female, in the very cold to in the hottest area. Before the development of smart textile; there were functional textiles, whatever they perform they are not active. They are passive. They are not designed to regulate themselves. They can be protective cloth but cannot be as the smart protective clothing. They can be water proof; however they are not vapour proof. All smart materials involve in transfer of energy sense for the stimuli to the actuator. These materials can process the data, which are sensed by the sensor and send the data to the actuators.

The smart garments in the form of PPE kit played a vital role like remote diagnosis and contact tracing in response to the current global COVID-19 outbreak. [1-5]

6. SMART TEXTILES AND WEARABLE TECHNOLOGIES

Opportunities offered in the fight against pandemics in relation to current COVID State: Smart textiles work as personal protective equipment (PPE), which works as a barrier between virus and human body keeping the comfort property in mind. One of them is including the chemical sensors in built in PPE kit, which would be able to detect viruses from the breath. It is also necessary to work on the development of "smart" PPE kit, which will not only protect but also detects and is capable of self-decontamination, durable and possibly biodegradable. In these circumstances telemedicine could be one of the best solutions to prevent the spreading of virus. In this case continuous monitoring of the physiological states of the patient is required. Therefore smart PPE kit would become most suitable one to monitor the body parameters. [1-5]

7. SMART TEXTILES IN A PLETHORA OF FIELDS

The development of smart textiles reaches like beyond imagination. The advancement in the field of smart textiles makes it possible to being the traditional textile sector, which start from hand loom to a level of high-technological industry. Moreover, it appears that this is only became possible by intense co-operation economic value and impact between people from various backgrounds and disciplines such as microelectronics, computer science, material science, polymer science, biotechnology, etc. Also more research needs to make it more convenient in our practical life. [1-5]

8. CONCLUSION

Smart textile are the combination of sensors and actuators, which sense if there is any change and at the same time react to those environmental conditions or stimuli, such as mechanical stress or strain, thermal, chemical, electrical resistance or capacitance, magnetic or other sources. The technology of SMART TEXTILES is not only limited to textile technology only but also became an integration of most of disciplines of applied sciences. Automobiles, robotics, aircrafts, medicine and surgery etc, may be a few example of the same. Smart textile either active or passive became inevitable parts of the military requirements i.e. camouflage effects. There is huge potential application of smart textile in the area of smart PPE kit to prevent spreading of virus by eliminate contact while monitoring psychological parameters.

REFERENCES

[1] Zhang, X.X. and Tao, X.M. (2001). "Smart textiles (1): Passive smart", *Textile Asia*, 32(6), 45–48.

[2] Zhang, X.X. and Tao, X.M. (2001). "Smart Textiles (3): Very Smart", *Textile Asia*, 2001, 32(8), 35–37.

[3] Smart Fibers, Fabrics and Clothing, Textile Institute, In X. Tao Ed., CRC Press, 2001.

[4] Das, S.C. and Chowdhury, N. (2014). "Smart Textiles—New Possibilities in Textile Engineering", IOSR *Journal of Polymer and Textile Engineering* (IOSR-JPTE), 1(1), 1–4.

[5] Das, S.C., Paul, D., Hasan, S.M.M, Chowdhury, N. and Nizam, M.E.H. (2013). "Smart Textiles—New Possibilities in Textile Engineering", *International Conference on Mechanical, Industrial and Materials Engineering 2013 (ICMIME2013).* https://www.academia.edu/12009400/Smart_Textiles_New_Possibilities_in_Textile_Engineering

Development of Sports Fabric Materials for the Improvisation of Performance and Wear Comforts for Sports Personnels

P. Rakshaa

Scholar, Kumaraguru College of Technology, Coimbatore, India

E-mail: rakshaanarayan01@gmail.com

ABSTRACT: *Depending on market demand, sportswear can be divided into four groups: performance sportswear, basic sportswear, sports casual wear, and sports fashion wear. Performance sportswear is a highly technology-oriented garment that improves performance through specific features. Competitive sports are sports that provide maximum physical activity in a short period of time. The comfort of highly active sportswear is becoming one of the most important quality standards as it can affect player performance. While performing sports activities, sweat tends to be produced which increases bacterial growth and releases an unpleasant odour. The odour may be due to the metabolic process of microorganisms. When the bacteria that cause such odours, are killed or suppressed, they are no longer metabolized and clothing and other items are odourless. In addition to the harmful effects of microorganisms on the human body, the demand for antibacterial clothing is increasing as consumers pay more and more attention to health and hygiene. These coatings minimize the risk of epidemics of infections and reduce hygiene problems. Preservation of odour-reducing sportswear and textiles requires bacterio static and slaughter treatments. These are the ultimate solutions to protect people from the bacteria that cause odours when they come into contact with the finished surface. Various chemical and natural means of combating odours and odour-causing organisms have recently been studied. The overview focuses on producing sports fabrics based on these properties.*

Keywords: Sports Fabric, Odor Causing Bacteria, Performance Fabrics, Odor Control.

1. INTRODUCTION

Sportswears are classified into Active, basic and performance sportswears. The requirements of active sportswear can be divided into two groups, i.e. functionality and aesthetics. Lightweight, low fluid resistance, high toughness, elasticity, thermoregulation, UV protection, breathability, sweat absorption, and release are major attributes for sportswears. [1] Competitive sports require maximum physical activity in a short period of time. The comfortable wearing of highly active sportswear is one of the most important quality standards as it can affect player performance. [2] Partial & for total suppression of bacterial growth can result in odour control. Therefore Sweat establishes a suitable environment for bacteria growths. Further metabolic process of microorganisms cause body odour. Sweats as well as moisture are management is an important issues in sports textiles. Textile materials transport water/liquid on the basis of capillary absorption principle.

Fibre moisture management during intense physical activity is an important issue that needs to be addressed to maximize the athlete's overall comfort. [1] *Propioni bacterium and Staphylococcus Albus* are common bacteria in sports field which cause infectious diseases. [3] Antibacterial textiles should be high-performance with respect to odour neutralization and prohibition of Skin problems. Bacteria and fungi can damage a wide variety of textile materials, leading to poor performance, spots due to ageing, odours, and infections on skin. Therefore, antibacterial textiles help reduce cross-infection and improve athlete performance by stopping the growth of microorganisms. [4] Demand for antibacterial clothing is increasing as consumers pay attention to health and hygiene in addition to the harmful effects of microorganisms on the human body. [5] Antibacterial-finished textiles play an important role in reducing the psychological discomfort associated with the growth of fungal microorganisms and the odour of textiles due to skin infections. They also form a strong barrier to the spread of antibiotic-resistant strains that cause nosocomial infections. [6]

Textiles, especially sports textiles, shoe insoles, or other odour-prone textiles, are treated with antibacterial substances to avoid unpleasant odours. Important antibacterial agents used in sports fabric applications include halogenated salicylic acid, anilides, aromatic halogen compounds, chlorinated diphenyl ethers (triclosan), benzoic acid esters, metal compounds (Ag, Zn, Cu, etc.), No. 4 Halogen compounds, triclosan, etc. The main objective of anti-bacterial sportswear is to protect the sportsperson from microorganisms that cause odours while also protecting the fibres from damage and breakage. Combined unconventional antibacterial technology works by allowing the technology to stick to the fabric. This protective shield technology does not dissolve or seep out and is not absorbed by microorganisms or consumers. Microbial membranes are physically ruptured by seams and electric shocks. [7] Organisms causing the smell can be treated with antibacterial agents, but deodorization and scenting is the process of trapping the substrate in scents/

essential oils. Pleasant scents can evoke a variety of specific emotions and emotions, including relaxation, cheerfulness, sensuality, happiness, and happiness, by stimulating the sensory pathways of the brain, especially the paleomammalian cortex. [8] The fabric of the garment fits snugly on the body and makes it easy to apply aromatherapy where it is needed. [9] Fragrance sophistication is recognized as a new area that has revolutionized the textile industry. Incorporating a variety of scents into the dough creates fragrant compounds that add value to the product and can affect the mood. [10] It is very important to extend the durability and consistency of the scent of the fabric to prolong its function. Aromas can also be incorporated into the dough by cross-linking the resin with essential oils. To inject the essential oil into the fabric, the perfume-containing compound can be paddle-cured directly onto the fabric using a squeegee roller, but the scent effect on the treated fabric lasts only after 23 washes. Based on the above referring to the issue of the effects of microorganisms on the sweating and odour of athletes and sports textiles, we have established the following objectives: In this study, antibacterial agents and aromatic oils were applied to knit materials and analyzed for antibacterial activity (EN ISO 20645 and fragrance test: Swiss standard SNV 195651). [11-13]

2. MATERIALS AND METHODS

2.1 Procurement of Fabric

Knit fabrics are widely used in everyday life, such as sportswear and outerwear. The fabric was made by using Fouquet circular knitting machine (36 knit systems, 30 needles per inch). The fabric samples produced were single jerseys and ribs weighing 170–200 g/m^2.

Table 1 shows the three types of fabric used in the survey.

Table 1: Fabric and Knit Types Used in the Research

S. No.	*Fabric*	*Knit Type*
1.	Lycra	Single jersey and Rib
2.	Cotton	Single jersey and Rib
3.	Poly cotton	Single jersey and Rib

2.2 Selection of Test Organisms

Test organisms such as Staphylococcus epidermidis and Propionibacterium were obtained from the local diagnostic laboratory, Coimbatore. Both cultures have been reported to be the causative odour-causing bacteria under heavy sweating conditions. All test cultures were grown and stored at 4°C before use. Each culture was freshly loaded into a nutrient solution and used for analysis.

2.3 Fragrance Oil

Aromatic tea tree oil was sourced commercially from Natural Oil Shops located in Coimbatore, Tamil Nadu and many parts of India.

2.4 Antimicrobial Compound

The antibacterial agent used in this study (trimethoprim-sulfamethoxazole) was purchased from a commercial pharmaceutical company. Use a UVVIS spectrophotometer to check the purity based on a specific wavelength.

2.5 Finishing Different Knitted Fabrics using Antimicrobial Drugs and Aroma Oil

In this study, we described how to treat antibiotics with hospital-based knit fabrics. The procedure was as described by Chun and Gamble for the reactive dye method. [14]

2.6 Reactive Dye Method

Synthesis of Reactive Antimicrobial drugs Using Beta-Cyclo-dextrin

The synthesis of reactive drug A (trimethoprim) was achieved by suspending 400 mg of trimethoprim (Ranbaxy) in 2 mL of deionized water in a 50°C ice bath. 2 ml of beta-cyclo-dextrin (Hi medium) was added to this suspension. The suspension was maintained at 50°C. while 1.0 N NaOH was added drop wise to completely dissolve.

Reactive drug B (sulfamethoxazole) synthesis was performed by suspending 200 mg of sulfamethoxazole (Sigma Chemical Co.) in 2 mL of deionized water in an ice bath at 50°C. Achieved. 2 ml of beta-cyclo-dextrin was added to this suspension. The suspension was maintained at 50°C. while 1.0 N NaOH was added drop wise to completely dissolve.

Dye-Exhaust Method to Bind Reactive Antimicrobial Agents to Cotton Fabric

An exhaust staining process was used to bond the reactants to the knit. Prepare dye bath by adding 0.2 ml of Triton X100 (Hi medium), 2 g of sodium sulfate and 2 ml of the reactive drug (A and B) to 100 ml of deionized water. Two 5 × 5 cm square knitted fabrics were dipped in a dyeing bath heated to 600°C. After 30 minutes of incubation, 1.0 g NaOH was mixed in 10 ml deionized water. After which, the temperature was set at 80°C. The knit was heated for an additional 30 minutes. The material is mildly washed with deionized water, heated to 80°C for 10 minutes and saved in a convection oven at 105°C y. A similar process was performed individually for all three types of fabrics (lycra, cotton and poly cot). [14]

Antimicrobial Activity of Finished Fabrics (EN ISO 20645 Test Method)

Fabrics treated with antibacterial agents were analyzed for antibacterial activity using the standard EN ISO 20645 test method. In short, the test piece (reactive drug-treated cloth) was cut into small pieces (diameter 20 mm). A sterile AATCC bacterio static agar plate was prepared. Cultures filled with loops (Staphylococcus Albus and Propionibacterium sp) were transferred by blotting the entire surface of the agar plate using a sterile 4 mm loop and covering the central area of the cell culture dish. The plates were incubated for 24 hours at 37°C. Inoculated panels were examined for failure to thrive along with the inoculum under the cloth, and separate inhibition zones beyond the edge of the cloth. Mean width of suppression zone around the sample (calculated in mm).

Fragrance Test: Swiss Standard SNV 195651

This test is performed according to the Swiss standard SNV 195651 and uses a modified sense of smell method to measure nasal irritation from fibres. Store a 20 cm^2 sample in an airtight desiccator (capacity approx. 2l) at 37°C and 50% relative humidity for 15 hours. With the help of saturated magnesium nitrate solution (about 100 ml), the humidity is adjusted. Under these conditions, at least five testers will open the desiccator for a short time to assess the intensity of the perceived odour. Odour intensity is assessed on a scale from 1 (odourless) to 5 (very strong odour). Once verified, the desiccator was closed again and stored under the specified conditions for at least an additional 15 minutes. After the test, the desiccator was left open at 37°C until the interior was odourless. And I updated the salt solution as needed. Based on the antibacterial and odour tests, at least one of the four best fabrics was finally selected.

3. RESULTS AND DISCUSSIONS

3.1 Antimicrobial Activity of Finished Fabrics (EN ISO 20645 Test Method)

In an analysis of antibacterial activity using standard test methods compliant with EN ISO 20645, single jersey fabrics of all two fabrics have more suppression zones against test organisms compared to ribbed fabrics. Lycra, cotton and poly cotton single jersey knits showed 29 mm and 27 mm antibacterial suppression zones. Staphylococcus albus and Propionibacterium sp are the respective skin pathogens. Relatively few restraint zones were found in both types of ribbed knits. Of the three finished fabrics, cotton ribs and cotton single jersey showed maximum restraint zones of 34 mm and 35 mm. Staphylococcus Albus and Propionibacterium sp, which are the respective skin pathogens. Against 33 mm and 35mm, Poly cotton subsequently showed good suppression zones of 34 mm and 34 mm. 33 mm and 32 mm for each test bacterium. Lycra showed relatively little activity during the analysis. This is evident in Table 2 and Figures 1 to 4. Interestingly, single jerseys for all three fabric types showed more activity than rib variations. This may be due to the slight washout effect of the single jersey fabric. The leaching effect was more pronounced during the incubation period and therefore the greatest inhibition was observed. The leaching effect can be affected by the way the fabric is knitted. Single jerseys are simpler to stitch than rib knits. Plain knits have more space than rib knits. For better antibacterial activity, more leaching was performed on the antimicrobial agent from the tissue.

The antibacterial activity of the reactive antibacterial agents obtained during this analysis was found to be important. The synergistic effect depends primarily on the mechanism of action of the drug on the bacterial cellular components. This has been shown to be synergistic if two different drugs have a similar mechanism of action. It is widely defined and discussed for two different groups of drugs called fluoro-quinolones and nitro imidazoles. Both drugs act on bacterial DNA and are intended to inhibit DNA synthesis and replication. (DNA gyrase and topoisomerase interact with IV Quinolones). (DNA gyrase was more sensitive to Gram-negative bacteria while topoisomerase IV was more sensitive to Gram-positive bacteria). Quinolone binding appears to cause changes in both DNA and topoisomerase, resulting in the formation of a three-component complex of quinolone, DNA, and DNA gyrase or topoisomerase IV. It that occurs separately from the DNA cleavage. It is characteristic of quinolone action. Inhibition of DNA synthesis by quinolones needs the target topoisomerase to have the ability to cleave DNA, and collisions between the replication fork. The reversible quinolone-DNA topoisomerase complex transform the DNA into an irreversible. [15] Therefore, the converted irreversible morphology destroys the double-stranded DNA produced and causes cell death. [16]

In this study, the mechanism of action of sulfamethoxazole inhibits beta-lactamase and prevents the destruction of trimethoprim. Therefore, to eradicate bacterial infections, sulfamethoxazole is given with trimethoprim to increase the activity of trimethoprim. This action of trimethoprim inhibits bacterial cell wall synthesis. Beta cyclodextrins contain cyclic oligosaccharides that form non-covalent complexes with drugs and alter their physicochemical properties. Potential sites for chemical modification are the primary and secondary hydroxyl groups of natural (α, β, γ) cyclodextrins. By incorporating these agents as a physical mixture into the drug delivery system, these agents bind covalently to the cross-linking agent, resulting in the constant release of the drug. Therefore, in combination, these drugs exhibit excellent antibacterial activity against odour-causing organisms.

Table 2: Antibacterial Challenge Test—EN ISO 20645 of Synergistic Drug Finished Fabrics against Skin-Associated Pathogens

S. No.	*Test Fabric*	*Zone of Inhibition (mm)*	
		Staphylococcus albus	*Propionibacterium sp*
1.	Lycra—Rib	23	26
2.	Lycra –Single Jersey	30	27
3.	Cotton—Rib	34	35
4.	Cotton—Single Jersey	35	33
5.	Poly Cotton—Rib	34	34
6.	Poly Cotton—Single Jersey	33	32

3.2 Fragrance Test

Anti-Odour Properties of the Drugs and Aromatic Tea Tree oil-finished fabrics.

This test is performed according to the Swiss standard SNV 195651 and uses a modified sense of smell method to measure nasal irritation from fibres. Store a 20 cm^2 sample in an airtight desiccator (capacity approx. 2 l) at 37°C and 50% relative humidity for 15 hours. Saturated magnesium nitrate solution (about 100 ml) is used to adjust the humidity. Under these conditions, at least five testers will open the desiccator for a short time to assess the perceived odour intensity. The depth of odour from 1 which indicated odourless to 5 which is considered to be a very strong odour is measured. After evaluation by the examiner, the desiccator was closed again and stored under the specified conditions for at least an additional 15 minutes. After the test, the desiccator was left open at 37°C until the interior was odourless. And I updated the salt solution as needed. Based on the antibacterial and odour tests, at least one of the four best fabrics was finally selected.

Table 3: Anti-Odour Properties of the Drugs and Aromatic Tea Tree Oil Finished Fabrics

S. No.	*Samples*	*Grades/Inference*
1.	Lycra—Rib	1
2.	Lycra –Single Jersey	2
3.	Cotton—Rib	1
4.	Cotton—Single Jersey	2
5.	Poly Cotton—Rib	1
6.	Poly Cotton—Single Jersey	2

Grades: 1-no odour, 2-not unpleasant, 3-slightly unpleasant, 4-unpleasant, 5-very unpleasant.

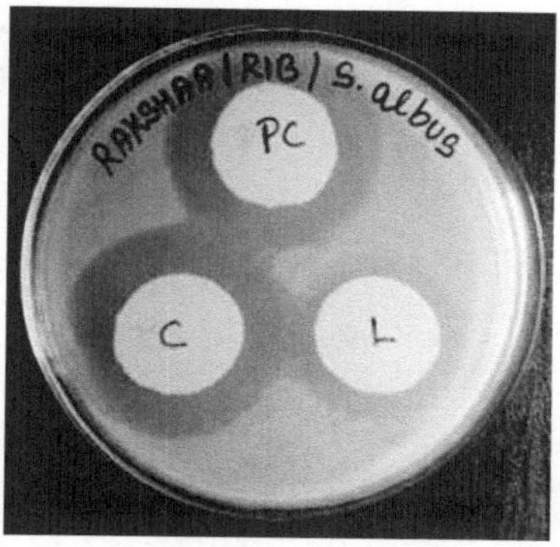

Fig. 1: Finished Fabrics (Jersey)
Staphylococcus albus

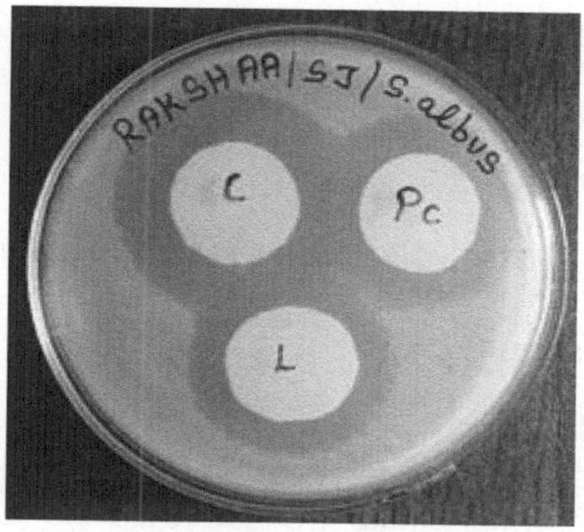

Fig. 2: Finished Fabrics (Rib)
Staphylococcus albus

C – Cotton, L – Lycra, PC – Poly cotton

Fig. 3: Finished Fabrics (Rib)
*Propionibacterium*sp

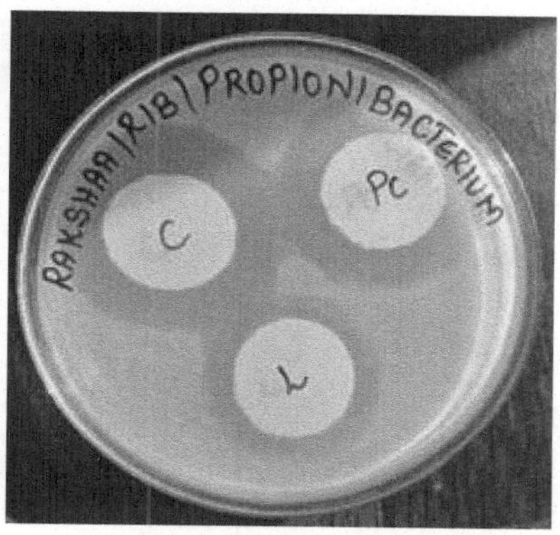

Fig. 4: Finished Fabrics (Rib)
*Propionibacterium*sp

C – Cotton, L – Lycra, PC – Poly cotton

4. CONCLUSION

This research focuses on the development of sportswear materials with enhanced antibacterial and aromatic properties. A total of four fabrics were selected for the study (lycra-rib, single jersey, cotton-rib, single jersey, poly cotton-rib, single jersey). The test fabric was finished with antibacterial agents (trimethoprim and sulfa methoxazole) using a dye-reactive process. Aroma tea tree oil was selected based on aroma tests to finish the dough. Therefore, it turns out that cotton single jersey knit fabric is more suitable for performance sports fabric. Cotton single jerseys have been found to be suitable for making sportswear fabrics, but other physical and mechanical tests need to be performed for future research. Optimization studies should also be conducted as future studies for the development and manufacture of effective high-performance fabrics with commercial quality and multifunctional properties.

4.1 Conflict of Interest

Authors declares no conflict of interest.

REFERENCES

[1] Sanjay, S., Rupali, S. and Ramakrishnan, R. (2004). "Waterproof Breathable Active Sportswear Fabrics". *Man-Made Textiles in India, 3*(5), 166–174.

[2] Manshahia, M. and Das, A. (2014). "High Active Sportswear—A Critical Review". *Indian Journal of Fibre & Textile Research, 39*, 441–449.

[3] Iqbal, M.N., Gallegos, A.M.A., Carrera, S.H., Parra, R. and Keshavarz, T. (2012). "Bacterial Cellulose: A Review". *The Indian Textile Journal,* 11(2), 66–72.

[4] Murphy, K.S. (2007). "Specialty Products for Processing of Textile Materials". *Colourage,* 29(6), 26.

[5] Yas. "Antibacterial Finishing Agents for Textiles". (2010). *Asian Textile Journal, 19* (5), 57.

[6] Gupta, S., Garg, N. and Saini, R. (2005). Text Book of Clothing Textiles and Laundry, Kalyani Publishers, New Delhi, 129–159.

[7] Prasad, A.K. (2007). "Novel Effects in Garment Processing and Value Added Finishes". *Journal of Textile Association, 68*(1), 39–42.

[8] Buchbauer, G. (1994). "Aromatherapy: Use of Fragrance and Essential oils as Edicaments". *Flavour and Fragrance Journal*, 9, 217–222.

[9] Salavica, S.M., Dejan, B. and Peter, S. (2005) "Microencapsulation in the Textile Industry". *Chemical Industry and Chemical Engineering Quarterly,* 12(1), 58–62.

[10] Kumar, D.V., Boopathi, N., Karthick, N. and Ramesh, P. (2012). "Aesthetic Finish for HomeTextile". *Indian Textile Journal,* 2(5), 22.

[11] West, A.J. and Hitchcock, A.K. (2014). "A Critical Review of Aroma Therapeutic Applications for Textiles". *Journal of Textile and Apparel Technology and Management,* 9(1), 48–61.

[12] "Textile Fabrics—Determination of Antibacterial Activity—Agar Diffusion Plate". *EN ISO: 20645,* 12–01.

[13] "Textiles; Determination Du Degagement D'odeurs Par Des Finissages". *Swiss Standards, 1968, SNV 195651.*

[14] Chun, D.T.W. and Gamble, G.R. (2007). "Textile Technology: Using the Reactive Dye Method to Covalently Attach Antibacterial Compounds to Cotton". *Journal of Cotton Science,* 11, 154–158.

[15] Hooper, D.C. (1999). "Mode of Action of Fluoroquinolones". *Drugs,* 58(2), 6–10.

[16] Laponogov, I., Pan, X.S., Veselkov, D.A., McAuley, K.E. and Fisher, L.M. (2010). "Structural Basis of Gate-DNA Breakage and Resealing by Type-II Topoisomerases". *PLOS ONE,* 5(6), 1–7.

Review on Hybrid Textile Reinforced Composites

Ashish Hulle*, U.J. Patil and Ravikumar Purohit

Department of Textiles, D.K.T.E. Society's Textile and Engineering Institute, Ichalkaranji, Maharashtra, India

E-mail: *ashishhulle@gmail.com

ABSTRACT: *Hybrid composites are synthesized by using various types of reinforcing agents. These are referred to as hybrids together. The goal is to mitigate the disadvantages of one fibre type or reinforcing agent while maintaining the benefits of the other. The hybridization process might result in synergetic effects or qualities that none of the elements has. Design flexibility is greater with hybrid composites than with non-hybrid composites. Hybrid composites provide design flexibility not just in terms of fibres and matrices, but also in terms of fibre orientations, stacking sequence, and preforms. Hybridization can be done in a range of ways. Hybridization can happen at three different stages: fibre, yarn, and fabric. The production, types of hybrid composites, the hybridization effect, performance properties, and diversified applications of hybrid textile reinforced composites are reviewed in this paper.*

Keywords: Hybridization, Interply, Polymer Composite, RoHM, Reinforced Material.

1. INTRODUCTION

Hybrid composites are those which consist of one kind of reinforcing agent/material that is used in a combination of two or more than two matrices (blends) or two or more reinforcing and or filling materials present in a single matrix. Sometimes both techniques are combined. Majority of textile reinforced hybrid composites possess of two or more continuous or discontinuous fibres in a single matrix. In textile reinforced hybrid composite with two or more fibre, reinforcement each fibre component supplements each other's short coming. By using the correct combination and form of the textile fibres, a cost-performance balance might be attained. The properties of the reinforcing fibres, their orientation aspect ratio extent of a blend and bonding & arrangement are the key factors for composites. It is essential to know the performance characteristics of the end application of the hybrid composite and accordingly select the fibre, form, configuration, etc. [1]

2. THE HYBRID EFFECT

Presence of hybrid structure improves is the apparent failure strain of a hybrid composite with low elongation (LE) fibre reinforcement compared to the failure strain of a non-hybrid composite reinforced with LE fibre. This explanation is shown in Fig(a).

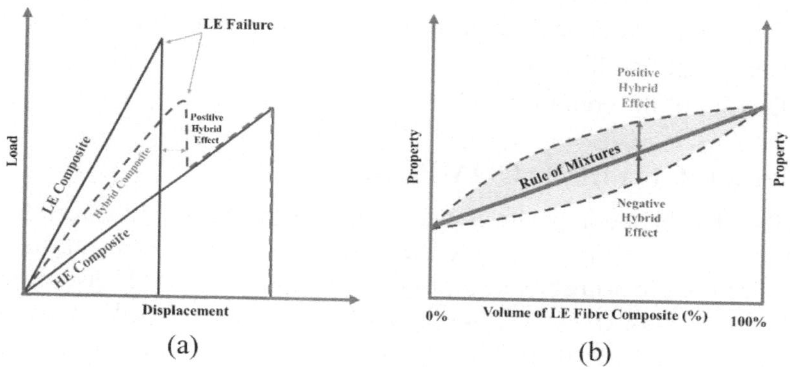

Fig. 1: Illustration of the Definitions of the Hybrid Effect

Another description of the hybrid by which one could get more information is a divergence from the simple rule of mixes. This is shown in Fig(b). The later description is more applicable to mechanical properties than failure strain. However, it is not applicable to all mechanical properties. [2]

3. RULE OF HYBRID MIXTURES (ROHMS) EQUATION

The mixing rule is applied to manufacture hybrid fibre reinforced polymer composites. The RoHMs is derived by considering the hybrid composite made by two single composite systems having no interaction between them. The elastic modulus of the hybrid composite (E_{hc}) can be calculated by using the RoHM equation:

$$E_{hc} = E_{c1}V_{c1} + E_{c2}V_{c2} \quad \dots (1)$$

Where,

V_{c1} = Relative hybrid volume fraction of first composite system.

V_{c2} = Relative hybrid volume fraction of second composite system.

$$V_{c1} + V_{c2} = 1 \quad \dots (2)$$

$$V_{c1} = \frac{V_{f1}}{V_f}, V_{c2} = \frac{V_{f2}}{V_f} \quad \dots (3)$$

$$V_f = V_{f1} + V_{f2} \text{ and } W_f = W_{f1} + W_{f2} \quad \dots (4)$$

$$V_{f1} = \rho_c \frac{W_{f1}}{\rho_{f1}}, V_{f2} = \rho_c \frac{W_{f2}}{\rho_{f2}} \qquad \dots (5)$$

Where,

V_f = Total reinforcement volume fraction.

W_f = Weight of the fibre.

V_{f1} = Comparative volume fraction of first reinforcement fibre.

V_{f2} = The relative volume fraction second reinforcement fibre.

ρ_f = Density of the fibre.

ρ_c = Density of the composite. [3, 4]

4. TYPES OF HYBRID COMPOSITES

Properties of hybrid composite is dependent on the form and orientation of fibre in the composite. According to the distribution of each element, they can be classified is three primary categories of (1) continuous and (2) discontinuous fibre hybrid composites viz. (a) Interlaminated, (b) Intermingled, and (c) Intra yarn or Intra ply hybrid composites. They are illustrated in Table 1.

Table 1: Types of Hybrid Composites

Form and Alignment	*Interlaminated*	*Intermingled*	*Intrayarn or Intraply*
1. Continuous (Aligned)	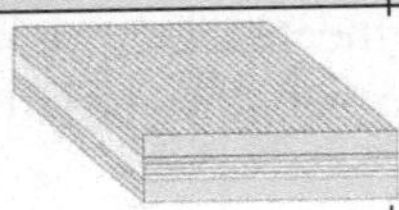	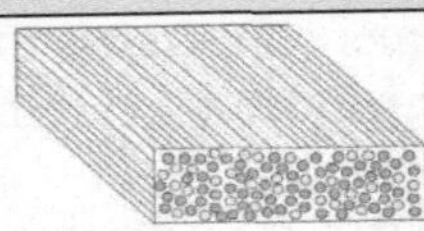	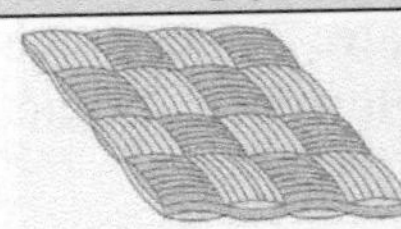
2(i) Discontinuous (Random)	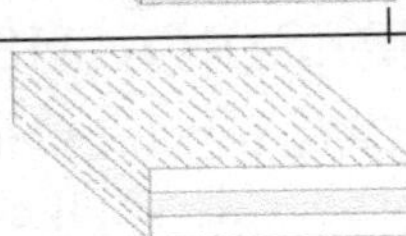	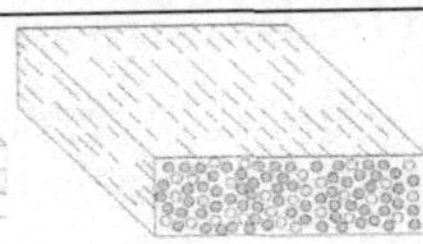	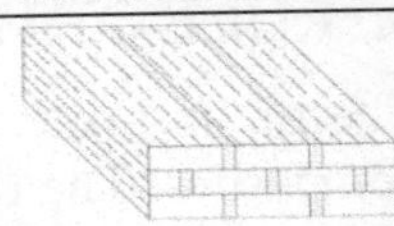
2(ii) Discontinuous (Aligned)	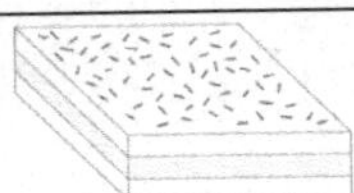	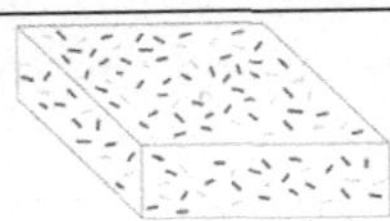	

Hybridization is done at the laminate level in interlaminated hybrids. Here, stacking of plies is made with different constituents. Bundles of different fibres are mixed or blended or different yarns are woven together respectively in intraply or intrayarn hybrids. Different types of components or constituents are mixed intensively to achieve hybridization within that ply for intermingled formation. [5] An interply hybrid laminate is made up of distinct fibre types in different laminas, whereas an intraply hybrid laminate is made up of two or more different fibre types interleaved in the same lamina. In most cases, the same matrix is employed throughout the laminate to establish a coherent interlaminar link between the laminas. [6]

5. PRODUCTION

The first step in the manufacturing of hybrid composite is the manufacturing of hybrid textiles and hybrid textiles can later be used in multiple configurations to create hybrid composites. With hybrid textiles, hybridization can be achieved within ply as well as between plies. The material as well as form of reinforcement and matrix, weave pattern in reinforcement, sequence of ply stacking, the number of layers or ply, fabric-matrix bonding adhesion, and processing conditions, plays a vital role in hybrid textile composites formation. There are multiple ways to achieve hybridization. When reinforcement is textiles, hybridization can be done at the fibre, yarn as well as fabric stages.

At the fibre stage, two fibres can be blended and formed into a web and that web can later be mixed in the polymer matrix to form the composites. Yarns made with two distinct fibres can be commingled together to form hybrid yams. These hybrid yarns are used to manufacture unidirectional or multidirectional prepregs. Hybrid fabrics can be woven by using different materials in warp and weft direction with various material and various constructional parameters. Not only yarns and fabrics but also nonwoven structures as well as braided structures can be hybridized for making composites. There are infinite possibilities of making hybrid textiles that could deliver a tailor-made performance in the hybrid composites. [7, 8]

After manufacturing the hybrid textile reinforcement, composites can be made using various ways that include hand lay, resin transfer moulding, compression moulding, pultrusion, injection moulding etc. The selection of manufacturing techniques depends on the application of composite. The right selection of composite processing settings can result in significant property improvements. [9]

6. PERFORMANCE PROPERTIES

Hybrid composites are mainly manufactured to augment the performance characteristics of composite with low manufacturing costs. According to many investigators, the tensile modulus of hybrid composites obeys a linear rule of mixtures. Hybrid composites show a 10–50% hybrid effect for failure strain. Ply layup configuration significantly influence the flex strength of hybrid composites. In hybrid composites by changing the material type of every ply and ply angle, the mechanical properties can be engineered. Complex lay makes it difficult to interpret the flex characteristics of hybrid composites. Simple rule of mixtures is applicable to tensile moduli. The traditional laminate theory is utilized to predict flexural moduli. One of the key reasons for fibre hybridization is toughening of the material. Impact resistance of composites is significantly affected by composite toughness. For many composite applications, fatigue resistance is an important performance characteristic. They composites demonstrate improved fatigue resistance than normal composites. LE fibres, as well as high elongation (HE) fibres, can be used to synthesize hybrid composite. HE fibres act as a crack stopper. The degree of dispersion was a critical criterion for all the qualities studied. Increased dispersion, in general, leads to bigger hybrid effects and higher performance. For complex loading conditions, different researchers have given conflicting opinions. [2]

7. APPLICATIONS

Recently, hybridised composites were used as part of the trough for collecting solar energy. Hybrid composites made with metal, metallic oxides and carbon-based fillers can be used for electromagnetic shielding and absorption. Synthetic fibres, such as aramid fibres, carbon and glass have been extensively used as reinforcement of resins to study their applicability in armour systems. Recently, natural fibres, combined with synthetic fibres, have drawn considerable interest for ballistic applications. [9] the most common use for hybrid composites is in commercial aeroplanes. Because composite structures require fewer riveted joints, they are lighter, which reduces fuel consumption and emissions while also improving aerodynamic efficiency and lowering manufacturing costs. [10]

Construction industry also unitized hybrid composites to make panels for false ceiling and partition boards, roof tiles, floor, wall, window door frames, etc. Automobile and railway body parts, coach interior, boat, gears etc. are also fabricated by using hybrid composites. Hybrid composites can also be used for making storage devices like post-boxes, grain storage silos, biogas containers, etc. They are used in making components for many electric devices. Apart from this hybrid composites find their uses in lampshades, suitcases, helmets, furniture etc. [3]

8. RECENT TRENDS IN FIBREHYBRIDIZATION

Initially, the focus of hybrid composites researchers was on lowering the cost of the hybrid composites. Later the focus was shifted towards manufacturing the composites structures with high failure strain for LE fibres. Though cost reduction is still a key objective behind the manufacturing of hybrids, several efforts have been made for a better balance of distinct material qualities or properties that are not present in the components. As a result, in recent times investigators are working on natural fibre hybrids, ductile fibre hybrids, pseudo-ductility, complex textile metal hybrids. There is a lot to explore in this area. [2]

9. CONCLUSION

Both academics and industry are becoming more interested in hybrid composites. A deeper knowledge of how reinforcing fibres work in composite materials allows for the design and manufacture of innovative textile-based composites for a range of applications. To fully realise the potential of metallic and polymer fibres, more study is required.

REFERENCES

[1] Jawaid, M. and Abdul Khalil, H.P.S. (2011). "Cellulosic/Synthetic Fibre Reinforced Polymer Hybrid Composites: A review." *Carbohydrate Polymers*, 86(1), 1–18. https://doi.org/10.1016/j.carbpol.2011.04.043

[2] Swolfs, Y., Gorbatikh, L. and Verpoest, I. (2014). "Fibre Hybridisation in Polymer Composites: A Review". Composites Part A: Applied Science and Manufacturing, 67, 181–200. https://doi.org/10.1016/j.compositesa.2014.08.027

[3] Sathishkumar, T.P., Naveen, J. and Satheeshkumar, S. (2014). "Hybrid Fiber Reinforced Polymer Composites—A Review". *Journal of Reinforced Plastics and Composites*, 33(5), 454–471. https://doi.org/10.1177/0731684413516393

[4] Venkateshwaran, N., Elayaperumal, A. and Sathiya, G.K. (2012). "Prediction of Tensile Properties of Hybrid-Natural Fiber Composites". Composites Part B, 43, 793–796. https://doi.org/10.1016/j.compositesb.2011.08.023

[5] Yu, H., Longana, M.L., Jalalvand, M., Wisnom, M.R. and Potter, K.D. (2015). "Pseudo-Ductility in Intermingled Carbon/Glass Hybrid Composites with Highly Aligned Discontinuous Fibres". Composites Part A: Applied Science and Manufacturing, 73, 35–44. https://doi.org/10.1016/J.COMPOSITESA.2015.02.014

[6] Mallick, P.K. (2007). Fiber-Reinforced Composites: Materials, Manufacturing, and Design. CRC Press. https://doi.org/10.1201/9781420005981.

[7] Mallick, P.K. (1997). Composites Engineering Handbook, CRC Press.

[8] Priyanka, P., Dixit, A. and Mali, H.S. (2017). "High-Strength Hybrid Textile Composites with Carbon, Kevlar, and E-Glass Fibers for Impact-Resistant Structures—A Review". *Mechanics of Composite Materials,* 53(5), 685–704. *https://doi.org/10.1007/S11029-017-9696-2*

[9] Alagirusamy, R., Fangueiro, R., Ogale, V. and Padaki, N. (2006). "Hybrid Yarns and Textile Performing for Thermoplastic Composites". *Textile Progress*, *38*(4), 1–71. *https://doi.org/10.1533/tepr.2006.0004*

[10] Balasubramanian, K., Purty, S., Khan, R., Ahmad, N. and Kayyum, M. (2019). "A Review on Recent Application of Hybrid Composites". *International Journal of Management, Technology and Engineering*, *9*(1), 193–201.

Development of Electro-conductive Fabric by Combination of Chemical Polymerization and Electrochemical Polymerization of Pyrrole

Syamal Maiti

Man-Made Textiles Research Association, Ring Road, Surat, India
E-mail: director@mantrasurat.org

ABSTRACT: *This work explains the effect of developing flexible electrically-conductive non-metallic textile fabric by two different processes such as in situ chemical and electrochemical polymerization. It has been reported the process of in situ chemical and electrochemical polymerization of pyrrole onto polyester fabric including process parameters and surface characteristics. The performance characteristics of the fabrics are expressed in terms of surface resistivity and surface roughness. It seems to be interesting to note that lowest the resistivity is achieved in case of electrochemical polymerization in spite of higher surface roughness. But between the electrochemical polymerization, low surface resistivity was achieved by lower surface roughness. There is change in resistivity observed as the in situ electrochemical polymerization process modified with graphene. Fabrics support electronic circuits prepared on its surface.*

Keywords: Pyrrole, Polyester Fabric, Chemical Polymerization, Surface Roughness, Electrochemical Polymerization, Surface Resistivity, Graphene.

1. INTRODUCTION

Electro-conductive polymers exhibit very good electro-conductivity, but the applications are limited on fabric because of low processability due to their inherent chemical structure. The stiffness which is caused by the high degree of conjugation along the polymer chain makes it difficult to process. In spite of limitations, researchers have developed electro-conductive fibers, yarns, fabrics, and other flexible products utilizing conductive polymers.

The use of electro-conductive textile materials is increasing in the area of electrostatic discharge garments, flexible sensing, heat generation, electromagnetic shielding, military applications, camouflage corrosion protection, and stealth material. [1-5]

In this work, an attempt have been made to develop the electro-conductive fabric in a two-step process as *in situ* chemical and modified electrochemical polymerization of pyrrole in combination with graphene and polyester fabric has been taken as base material. In addition, it was intended to study the effect of electrical conductivity, surface roughness and electromechanical behaviour the process onto fabric. It was also made an attempt to print electronic circuit on the fabric.

2. EXPERIMENTAL

In this work, hydrolysed polyester fabrics have been taken as base material. The chemicals used were laboratory grade sodium carbonate, sodium hydroxide, non-ionic detergent lissapol N, pyrrole, ferric chloride, and p-toluene sulphonic acid. The electro-conductive fabric was prepared by a two-step process a) *in-situ* chemical and b) *in-situ* electrochemical polymerization of pyrrole combination with graphene onto polyester fabric.

2.1 *In situ* Chemical Polymerization

In-situ chemical polymerization was carried out on the initially hydrolysed Polyester fabric. The hydrolysed fabric was treated with pyrrole 0.5 M pyrrole solution in water at 25°C for 20 min. Followed by the pyrrole-enriched fabric was immersed into 0.75 M ferric chloride solution. Ferric chloride initiates polymerization and the process was continued for 10 min at 5°C.

2.2 *In situ* Electrochemical Polymerization

The *in-situ* electrochemical polymerization of pyrrole was done on already chemically polymerized fabric. This process was performed using a regulated DC power supply. The solution contains pyrrole, p-toluene sulphonic acid used for polymerization. The chemically polymerized fabric was fixed on anode surface. In other experiment, the electronic circuit was prepared by combination of chemical and electrochemical polymerization of pyrrole on the surface of polyester fabric.

2.3 Characterization

The surface resistivity of the fabric was determined using ASTM standard D-257 by measuring the surface resistance with digital multimeter. The surface characteristics of the fabric were studied by scanning electron microscopy (Zeiss EVO 50 scanning electron microscopy system). The surface roughness of the fabric was measured by Talysurf Intra (Taylor Hobson Precision, UK) using stylus.

3. RESULTS AND DISCUSSIONS

3.1 Effect of Polymerization on Surface Resistivity

After chemical polymerization of pyrrole on hydrolyzed polyester fabric, a little amount of polypyrrole is deposited on the surface. As a result polyester fabric became slightly conductive. But as a result of electrochemical polymerization, the amount of polymer deposited is enormous as shown in Figure 1. It leads to decrease the resistivity of the fabric enormously.

Hydrolyse | Electrochemically Polymerized

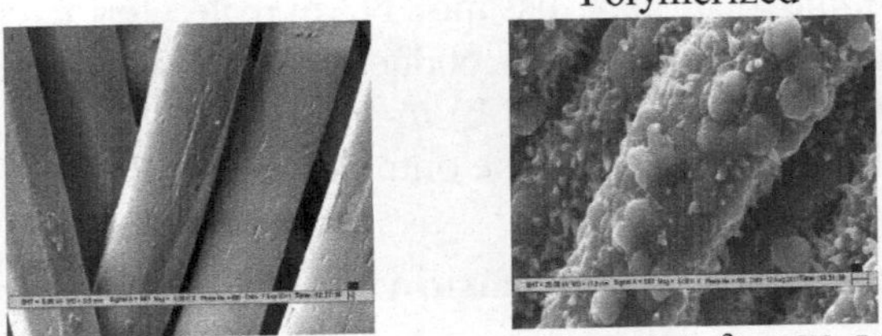

Add-on (g/m^2): 0.0 | Add-on (g/m^2): 170.54

Resistivity (Ω): >10^{14} | Resistivity (Ω): 53.66

Fig. 1: Scanning Electron Microscopic Micrograph of Electro-Conductive Fabrics

It could be noted that the amount of polymer deposited on fabric does not correlate with the surface resistivity. This may be due to relatively high rough surface developed at higher deposition on the fabric. The surface roughness and surface resistivity correlate well, which can be proved by the surface roughness data (Table 1).

Table 1: Results of Surface Roughness and Surface Resistivity

Material	*Weight Per Unit Area (g/m^2)*	*Surface Roughness (μm)*	*Surface Resistivity (Ω)*
Fabric	34.90	14.5	46.26
	132.73	12.4	14.19
	254.63	31.0	181.21

3.2 Effect of Polymerization on Surface Printed Circuit

The flexible electronic circuit made on the surface of polyester fabric has been shown in Figure 2a (open circuit) and Figure 2b (closed circuit). The

LED glows when the circuit is closed by touching the fingertip between two ends of the printed area as the circuit became closed by touching.

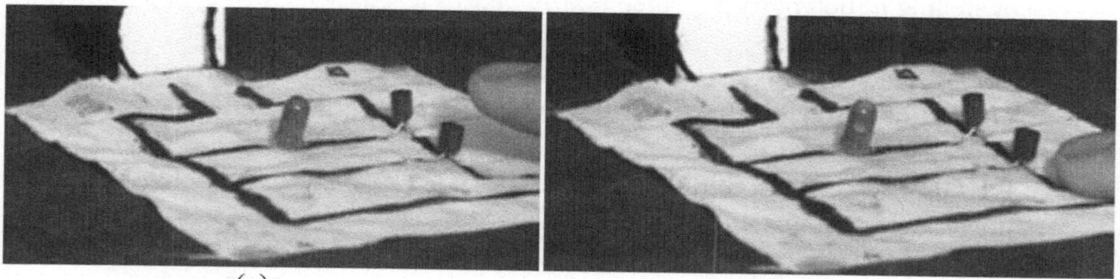

(a) (b)

Fig. 2: Photograph of Flexible Electronic Circuit a) Open Circuit b) Closed Circuit

3.3 Voltage-Current Characteristic of Electro-Conductive Fabric

The voltage-current characteristics of electro-conductive fabric have been studied by applying voltage through the fabric and corresponding current have been recorded. The results of experiments are shown in Figure 3. It may be noted that a non-ohmic increase in current has been observed as voltage was applied across the fabric sample strip.

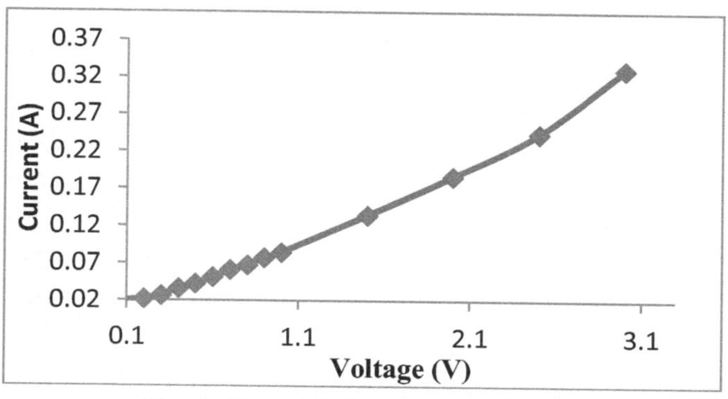

Fig. 3: Plot of Voltage versus Current

4. CONCLUSIONS

The electrochemically polymerized fabric has shown low surface resistivity compare to chemically polymerized fabric as deposition is moderate. Also, lower surface resistivity has been observed at higher add-on but lower surface roughness. It has been observed that the electro-chemical polymerization is a better process to get less resistivity but one must have control to keep the surface roughness as low as possible. The electro-conductive fabric supports the electronic circuits prepared on it.

REFERENCES

[1] Kincal, D., Kumar, A., Child, A.D. and Reynolds, J.R. (1998). "Conductivity switching in polypyrrole-coated textile fabrics as gas sensors". *Synthetic Metals*, (92), 53–56.

[2] Boutrois, J.P., Jolly, R. and Petrescu, C. (2009). "Process of polypyrrole deposit on textile- Product characteristics and applications". *Synthetic Metals*, (85), 1405–1406.

[3] Kraljic, M., Mandic, Z. and Duic, L. (2003). "Inhibition of steel corrosion by polyaniline coatings". *Corrosion Science,* 45(1), 181–198.

[4] Kuhn, H.H., Child, A.D. and Kimbrell, W.C. (1995). "Toward real applications of conductive polymers". *Synthetic Metals,* 71(1–3), 2139–2142.

[5] Hakansson, E., Kaynak, A., Lin, T., Nahavandi, S., Jones, T. and Hu, E. (2013). "Characterization of conducting polymer coated synthetic fabrics for heat generation". *Synthetic Metals,* 144(1), 21–28.

Painting with Natural Dyes on Textiles

Silpinwita Das and Sankar Roy Maulik*
Department of Silpa-Sadana, Visva-Bharati (A Central) University, West Bengal, India
E-mail: *s_r_moulik@yahoo.co.in

ABSTRACT: *Natural dyes used for textiles are primarily derived from renewable vegetable resources, viz. barks, roots, flowers, fruits, leaves, seeds etc. Dyeing and printing/painting with natural dyes on textile substrates were done dating back to the dawn of human civilisation to enhance the aesthetic appeal. India has a wide variety of linguistics, landscapes, music, and dance forms, perfectly echoing the diverse history of natural dyes in textiles throughout her stretch. In the present era of ecological concern, there is a worldwide interest in dyeing and painting/printing fabrics using natural dyes. Nowadays, consumers' demand is the prime importance for the survival of any business. Indian block printed, resist-dyed, hand-painted, handwoven textiles were praised and valued worldwide for over a millennium. The present paper explores several aspects of painting with natural dyes on fabrics. A brief discussion on the profound history of painting in the Indian subcontinent has also been addressed. Few textile samples have been developed using Terminalia chebula, Rubia tinctoria, Acacia catechu, Curcuma longa, Indigofera tinctoria etc., as the primary colourant with the addition of different innocuous inorganic salts viz. ferrous sulphate, copper sulphate, potassium aluminium sulphate as mordants following the traditional process of dyeing and painting largely. The final products can also be utilised as attire, accessories, furnishing fabric apart from wall hanging as they show good colour fastness properties.*

Keywords: Ecological, Dyeing, Natural Dye, Painting, Textile.

1. INTRODUCTION

Painted textiles include a vast array of functional and decorated fabrics. It can be surface painting on stiffened cloth viz., Pattachitra, Pichhvai, Phar, Thanka etc. Also, it can be painted with an introduction of metal salts, i.e., mordants that can permanently fix the dyes with the fibres viz., Chintz, Mata nipanchedi, Kalamkari etc. Most of the natural dyes used for textiles are obtained from vegetable resources, i.e., barks, roots, flowers, fruits, leaves, seeds, and animal sources. Besides fixing the dyes with the fibres, the mordants also help produce different shades of colours from a specific dye source. The use of alum for fixing the colour has been reported in the ninth century, although evidence suggests that it may have occurred at a much earlier date. [1] Not only can the mordant, but even the composition of the water also affects the colours. [2] Dyeing and printing with natural dyes on textile substrates were practised dates back to the dawn of human civilization to enhance the aesthetic appeal and cure specific diseases due to its proximity to the skin. [3, 4] The latter kind of fabric is often called Ayurvastra and is used in apparel to floor mats, towels, to medicinal clothes. [5] The fragments of madder dyed cotton attached to a silver vase discovered in the excavation of Mohenjodaro evident the profound knowledge over natural dyes with mordants even in 2500 BC. In the 1st millennium BC (1st to 10th C BC), Panini wrote about textiles known by their colours, such as Nila, Lohitaka, Laksha, Kalaka. [6] In Samyutta-Nikaya, Buddha mentioned resin, lac, turmeric, indigo, madder used as dyes or colours by the dyers or painters. [7] The Dutch and Portuguese traders introduced India's block printed, resist-dyed or painted textiles (Chintz) to Europe in the 17th century. [8] Later, in the 17th C, the British's establishment of the East India Company initiated a thorough new chapter in the history of Indian textiles. The land which once dominated and influenced the world trade for over a millennium was affected by the influx of cheap mill clothes. Then in the 19th C. with the advent of synthetic dyes throughout the world, the real revolution started. Many countries all over the globe gradually adopted synthetic dyes. This is cheap and needs a small quantity than any natural colour to work with, and they are easy to apply. Those dyes are extracted from fossil resources, coal tar and oil, andare harmful to the planet. On the contrary, there are many advantages of using natural dyes on textiles: i) it is biodegradable and eco-friendly, ii) safe for skin, food and children, iii) the waste after dyeing can be used as fertilisers (indigo and myrobalan etc.), iv) unique shades can be achieved, etc. Despite these advantages, natural dyes have certain inherent limitations, which are mainly responsible for declining this traditional ancient art. They are: i) despite its soft and varied tones, reproducing and standardisation of recipe for a particular shade is challenging, ii) it is expensive as it has comparatively low colour yields, iii) the colour also may change if it is exposed to light, sweat, air etc., iv) the dye bath consisting of various salts may create a severe disposal problem. Today there is a market for such textiles preferred by environmentally conscious consumers. The current natural dyestuff requirement from the industry is about 3 million tons. [9] This customer-driven massive industrial production is becoming a significant economic and cultural issue and a potential threat to some of the biodiversity that needs to be preserved. Also, consciousness and understanding over natural dyes can be developed to differentiate between the true natural colours and fraudulent imitations based on synthetic dyes of it. Some painted samples using natural colour on cotton in the presence and absence of different mordanting agents have been developed in this research work to add value. The colour fastness properties viz. light, rubbing and wash of those painted textiles have also been tested following standard procedure.

2. MATERIALS AND METHODS

2.1 Material

Terminalia chebula, *Rubia cordifolia*, *Acacia catechu*, *Punica granatum*, *Indigofera tinctoria* have been used along with the fermented solution of scrap iron and molasses. Potash alum and copper sulphate have been used as mordants. Gum Acacia has been used as a thickener. All those materials and fabrics have been sourced from a local market near Burra Bazar, Kolkata. Indigo is sourced from KMA Exports, Tamil Nadu. Normal tap water has been used throughout the dyeing and washing.

2.2 Methods

2.2.1 *Scouring of Cotton*

To remove natural and added impurities before dyeing/painting, scouring of cotton fabric was performed. The scouring of cotton fabric was done in the presence of wood/leaves ash (20%), fresh cow manure (20%) and calcium hydroxide (5%). All measurements were based on the dry weight of the fabric. In this traditional tub method, the alkaline solution was made by putting the dry ash material into hot water and bringing it to a boil. Next, cow manure and calcium hydroxide were added and stirred thoroughly. It was then filtered, and more water was added with this aqueous solution in which the fabric was simmered for approximately 30 minutes and rested overnight. The material needs to beat the next day rigorously before rinsing in clean water and finally drying in the sun. This is called half-bleach. For Full bleach, the fabric undergoes the same procedure. But instead of the washing stage, the materialgets steamed. It was laid on the ground in direct sun and occasionally sprinkled with manure water when the fabric was near dry but not thoroughly dried. This process is repeated for 3 to 4 hours. For the present study, the half-bleach method was adopted.

2.2.2 *Tanning*

The fabric was soaked overnight in crushed myrobolan (10–20%) solution and dried in the sun. It is a necessary step for the black and red colour processing.

2.2.3 *Extraction of the Natural Dye*

The aqueous solution of natural dyes was prepared by crushing the crude materials, then soaking and boiling them in water. After boiling, it was cooled and filtered. An average of 50 g of crushed source (except Indigo) was taken for extraction to make 100 ml dye solution.

2.2.4 *Painting Process*

In an earthen pot, 500 g scrap iron; 1kg molasses and 5 L water were kept for 20–25 days. It was then filtered and applied on the myrobalan treated cloth to achieve black. As myrobalan has high tannin content, it doesn't require any other mordant. An aqueous solution of alum (4%–5%) was applied only to the required areas. The fabric was washed thoroughly in flowing water and dried in the sun. Then dyeing was carried out with *Rubia cordifolia* for 20 minutes at 60–70°C. Only the alum treated areas will arrest the anthraquinone derivatives from the dyebath resulting red. After washing, it was dried and was ready for the further application of the rest of the colours. Gum Acacia as thickener and potash alum as the fixing agent was added with the dye solution. Copper sulphate was added only for *Acacia catechu*. Preparing indigo powder ready to paint is a different process than the other colours. 20% indigo, 40% calcium hydroxide, 60% reducing agent was taken to prepare a vat. With the added thickener, the fermented indigo is applied to the fabric. It has to be repeated several times to get the desired shade. Different shades and tones of colours can be achieved with the permutation and combination of the process mentioned above. The fabric needs to be washed in cold tap water and dried finally.

2.2.5 *Assessment of Colour Fastness*

Colour fastness to washing of selected fabrics painted with natural dyes was assessed in accordance with a method prescribed in ISO 105-C06:2010 employing 5 g/l soap solution at a temperature of 50°C for 45 minutes at fabric to liquor ratio of 1:50. Colour fastness to the light was assessed following a method prescribed in ISO 105-B02. Colour fastness to rubbing was determined employing a crock meter following the method prescribed in ISO 105-X12:2001. [2]

3. RESULTS AND DISCUSSIONS

Figures 1 shows different painted textiles with natural dyes. The colour fastness to washing and rubbing of those painted textiles was found very good, where as the light fastness rating was also good.

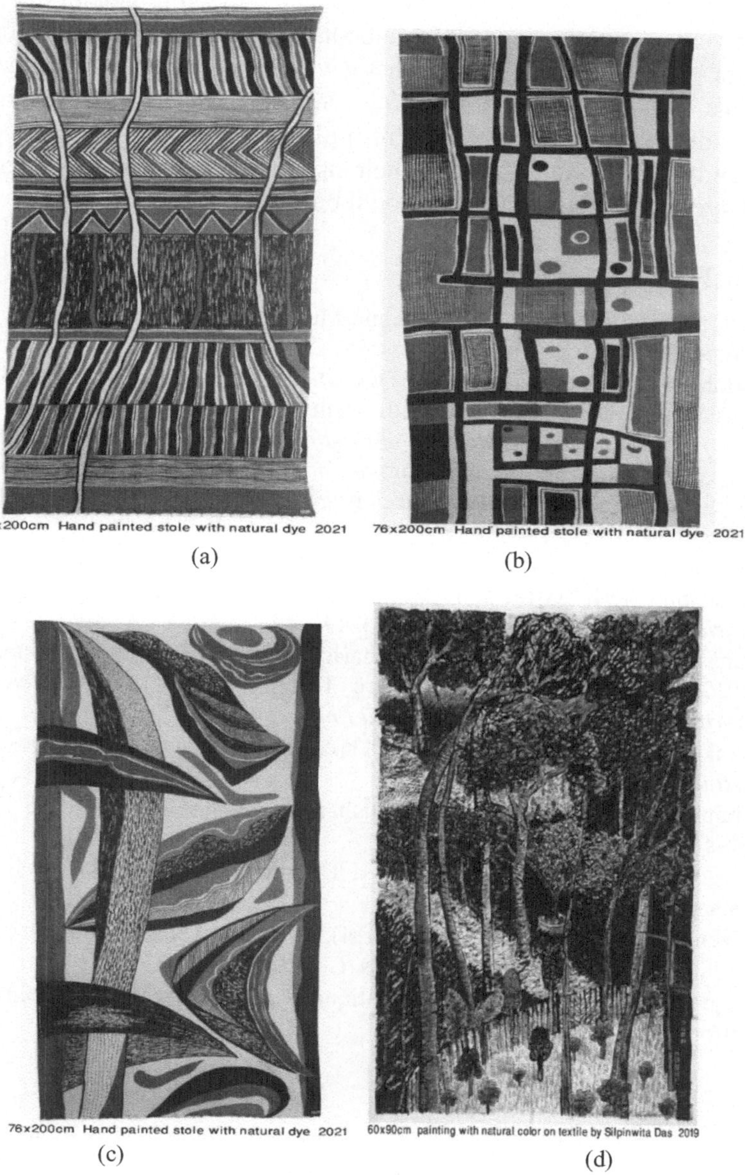

Fig. 1: (a)–(c) Stole & (d) Painted Textile with natural Dyes

4. CONCLUSIONS

India has a rich tradition of painting with colourants obtained from vegetable resources. Painting with natural dyes contributes to value-added textiles. Through this process of traditional painting, one can give a crafty look to the cotton fabrics used for apparel.Natural dyes are emerging globally because of the growing consumer awareness about environmental conservation and control of pollution. Plants sources that have the potential to be used as textile colourants can be explored further for commercialization. Efforts should be made to identify the raw materials from plant sources and to standardize the recipe and dyeing conditions for their use. There is currently demand for natural dyes in a niche market that could be explored.

REFERENCES

[1] Roy, M. (1978). "Dyes in Ancient and Medieval India". *Indian Journal History of Sciences*, 13(2), 84–85.

[2] Mridha, A., Mukhopadhyay, A., Das, D., Sarkar, B. and Roy, M. S. (2021). "Sustainable Dyeing of Silk with Natural Dyes Using Natural Sources of Water". *Journal of the Institution of Engineers* (India): Series E, 102(1), 61–74.

[3] Roy, M.S., Bhowmik. L. and Agarwal, K. (2014). "Batik on Handloom Cotton Fabric with Natural Dye". *Indian Journal Traditional Knowledge*, 13(4), 788–794.

[4] Chandrasekaran, K., Ramachandran, T. and Vigneswaran, C. (2012). "Effects of Medicinal Herb Extracts Treated Garments on Selected Diseases". *Indian Journal Traditional Knowledge*, 11(3), 493–498.

[5] Rangari, N.T., Kalyankar, T.M., Mahajan, A.A., Lendhe, P.R. and Puranik, P.K. (2012)."Ayurvastra: Herbal Couture Technology in Textile". *International Journal of Research in Ayurveda and Pharmacy*, 3(5), 735.

[6] Jayakar, P. (1980). "Indian Textiles Through the Centuries". *Treasures of Indian Textiles*, Marg Publications, 59.

[7] Kramrisch, S. (1928). "The Vishnudharmottara Part III". Calcutta University Press, 15.

[8] Varadarajan, L. (1979). "Towards a Definition of Kalamkari, a Homage to Kalamkari", *Marg Publication*, 19–21.

[9] Saxena, S. and Raja, A.S.M. (2014). "Natural Dyes: Sources, Chemistry, Application and Sustainability" Issues, *Google Scholar*, 2.

[10] Wisnaik, J. (2004). " Dyes from Antiquity to Synthesis", *Indian Journal of the History of Science*, 39(1), 76.

Seamless Garmenting: An Overview

Ajay Shankar Joshi, Tanveer Malik and Shyam Barhanpurkar
Shri Vaishnav Institute of Textile Technology, SVVV, Indore, India
E-mail: *ajayjoshi@svvv.edu.in

ABSTRACT: *Seamless clothing is very close to the human body just as second skin covers the body perfectly & comfortably, details about the seamless garments are discussed, such as its manufacturing techniques, machines used, advantages, and disadvantages. The common seamless products also described with their application. The micro fibers lycra, cotton yarn are the prominent raw materials for seamless garment. The fine filaments of nylon produced the softest and most sensual fabrics. Seamless clothing can be produced by the knitting technique through either a circular knitting machine or a flat (V-bed) knitting machine, with a greater variety of structures. The seamless garments knitted on a circular machine, which lead to minimal cutting and seam joining processes on one body tube and two sleeve tubes as well as the finished edges. Application of Seamless clothing in various sectors like medical textile, automotive, apparel, under garment, upholstery, socks, gloves, protective cloths etc. The advantages of Seamless clothing are cost saving, waste reducing, time saving, comfortable garment, flexibility in design etc. The seamless garment has some disadvantages also, that discussed. In this paper a detail explanation of its techniques, advantages, needs, disadvantages are discussed.*

Keywords: Seamless, Knitting, Garmenting, Comfortable, Micro-Fibre, Upholstery.

1. INTRODUCTION

The seamless garments are generally closed to human body skin just like considered as second skin. They are also very comfortable and perfectly fitted garment. They give elegant appearance in outer as well inner wear segments. The history of seamless garment is very old. The seamless technique was common in practice even in 4th & 5th century AD. The seamless technique was used for the manufacturing of socks in Egypt during Coptic period of time. In year of 1940, body shaped skirts was patented in USA by using seamless technique. The main objective of that production was reduction cost by the fleck age technique and enhanced the drape & fitting of skirts (Hiatt, 1988).

In 1965 automatic knitting machine produced shaped knitted components. In year of 1960, Shima Seike produced the gloves by tubular type knitting principle. They also developed shaped seamless knitted components (Rutt, 2003). Subsequently in 90's era seamless garments covered only 1% production of total knitted production. The market share figure has risen up to eleven percent and it is continuously rising. Swim suits, bra and knick all are the classical example of the concept of seamless garment. In modern era of machinery, seamless garment concept opened new sector of business. Now the seamless technique is competing with traditional garment manufacturing processes.

The seamless technique has become more attractive for manufactures due to the adverse consequence of cutting & sewing processes under the tradition processes. Few are narrated below:

1. Cutting & sewing both are labour intensive.
2. Chances of fabric damage due to continue penetration of needle during sewing.
3. Probability to alter the mechanical properties of fabric.
4. Human error.

The seamless technique is considered as simple technique. In tradition approach the cut-and-sew production process generated 40% wastage of fabric where as fully fashioned garment is produced by seamless technique through loop transferring process. The seamless techniques has unique feature to change the dimension of fabric by changing the numbers of wales though V bed flat machine. [4] The cutting process is eradicated in this process but it requires a post sewing or linking process.

Addition to that, there is the opportunity in seamless technique to create single knit reversible garments without added extra weight and bulk. Although these products are popular in the underwear sector, particularly for ladies' underwear, that covers markets share about 78%. Apart from this sector, they are also hold market share in another sector, like men's underwear (13%) outerwear (2%) to swimwear (2%), sportswear (4%) and sanitary products (1%). [1-6]

2. SEAMLESS TECHNIQUES

The Seamless garments produced by the successive knitting cycle on several different carriers (feeders) without cutting and sewing processes. Seamless flat knitting machines such as the Shima have capability to produce more than one knitted tubes on same time then that tubes are joined together on the machine. The complete seamless garments are also produced on circular machines with altering courses & wales. [7-9]

2.1 V-Bed Machine Knitting of Seamless Garments

As mentioned earlier, seamless garments produced by flat V-bed knitting machine with three separate tubular forms according to body shape, in that one wider tube for the upper body part and two narrower tubes for the sleeve parts. Tubular knitting completed on both needle beds but knitting is done alternately on front and back beds. [3]

On the flat knitting machine, seamless garments produced by the loop transfer process between front needle bed and back needle bed with different yarn carriers. In that arrangement different yarn carriers are assigned to make different tubes like for one upper body tube carrier number 2 is responsible whereas carrier number 2&3 are responsible for other two sleeve tubes are. The three-tube knitting process continues to the under arm portion of body, at that portion, sleeve parts are taken out from knitting zone. The remaining knitted tube for body part will knit tighter with other tubes and make one single tube at output zone. All the tubes are joined perfectly at the underarm points, shoulders and neck points:

1. The loop transfer process produced the Seamless garment very easily.
2. The loop transference process contains the uses of alternate needles and the single jersey tubes as well as double jersey tubes that are knitted to make seamless garments by this process. [5]

2.2 Wale Shaping

The principle involves knitting by alerting the number of Wales at internally within a flat piece of fabric/bra tube of knitting by keeping the number of courses same.

2.3 Course Shaping

The principle involves changing successively the length of the courses being knitted alternately. It can be described as knitting at differing number of loops on Wales.

3. APPLICATIONS

3.1 Gloves

The seamed gloves have problem of getting perfect shapes for cuff, palm, thumb and the four fingers areas. These problems are solved by complex knitting process, which produce entire gloves with closed fingers, thumb and elasticized wrist. In latest knitting machines, they have feature to knit the gloves with tuck loop.

3.2 Upper Body Garment

The seamless garment has unique characteristics to make garment according to the body shape through the successive numbers of knitting cycle. In the seamless techniques as cycle of knitting progresses then sleeves are to be merged to the body accordingly.

3.3 Upholstery

The stretch property is one of the important requirement for upholstery products. The seamless seat upholstery products have inherent features of stretch ability with comfortable feel, better fitting characteristics. They eliminated the ridges caused by the sewing proesss. The companies like Teknit, Courtaulds, General Motors Corporation, Lear Corporation etc are the major player in this market segment.

3.4 Medical Textiles

Now a day's medical textiles is one the emerging and growing sector especially in the COVID-19 pandemic situation. The seamless products have potential to use in medical sector.

3.5 Application in Automotive

The seamless products has potential to use in automotive sector. In the industrial fields like automotive and aerospace, a seamless fabric could be easily useful in a variety of belts, hoses and specialized covers. In this segment, there is scope for seamless products to make advanced products with different shapes as per the technical specifications.

3.6 Protective Clothing

The application of seamless garments is not only in fashion apparel market but also in life saving market sector. The seamless garments are applicable as protective garment.

4. RAW MATERIALS

Seamless garments are generally produced by the blend of the micro-fibre yarns, with lycra and cotton. The structure of Micro-fibre yarns comprises with fine filaments of nylon fibers that make the softest and most sensual fabrics. Unlike the regular nylon, micro-fibre is similar like regular cotton. The use of nylon fiber is dominating for hosiery products, but the use of DuPont's CoolMax proved that polyester fiber is also highly compatible with seamless technology.

Natural fibers give smooth look and soft touch. They have characteristic to reduce the pilling and they are useful in bodysuits, lace. A new arrival among seamless wear fibres is Sensil by Nilit. Sensil Arafelle is 100% nylon 6, 6 fibre. They re also compatible for jacquard knitting process.

5. ADVANTAGES OF SEAMLESS GARMENTS

5.1 Improved Aesthetic Value and Comfort

The appearance of a garment is enhanced, as there are no seams in the garment. Also there are no chances of seam puckering, mismatching of patterns and fitting related problem. Seamless garments consist of limited buttons, hooks, zippers. They have superior comfortably, smoothness and sleek fitness.

5.2 Cost Saving

The costs are involved in other processes like inspection of fabric, storing of fabric followed by spreading, cutting, stitching during seamed garments, which are eliminated in seamless technique. They have additional features like low labour cost, floor space, cost of machines and comparatively low consummation of power.

5.3 Waste Reduction

The fabric wastage generally occurs during spreading, cutting processes and subsequent stitching operation. The exclusion of these processes saves the yarn and fabrics.

5.4 Lower Lead-Time

There is no dependency on the fabric supplier or supplier of other stitching accessories, which is one of the causes of production delay. This gives the lower lead-time.

5.5 Flexibility

It provides flexibility to knit a garment with special features like moisture management in the garments. It provides more creative possibilities for knitwear designers. The other advantages are more constant product quality, better trimmability for finished edge lines, just-in time production and mass customization. [7-9]

6. DISADVANTAGES OF SEAMLESS GARMENTS

- The main constraint is to keep equal tension on each loop during fabric take down process.
- Problem in arrangement of alternate needle selection process that gives more open fabrics and comparatively less elastic than seamed garments.
- Skilled operators are required.
- The occurrence of knitting defects (particularly a hole), that damage the whole garment.
- The seamless garments are comparatively considered as costlier garment then conventional garments.

7. CONCLUSION

In the apparel market, the seamless technique is revolutionary technique because of its smoothness, comfortable fitting, invisibility and easy care properties. The world wide trends predict that seamless garments will be considered as most popular garments especially among the youth. The seamless garments have potential to cover fifty percentages of the industry's sales in coming ten years by exploring more technical trainings with value added facilities. That share would lead by adding unique product design features.

REFERENCES

[1] Hiatt, J.H. (1988). "The principles of knitting: Methods and techniques of hand knitting". *Simon and Schuster*, New York.

[2] Rutt, R.A. (2003). "History of hand knitting". *Interweave Press*, Loveland, CO.

[3] Spencer, D.J. (1989). "Knitting Technology: a Comprehensive Handbook and Practical Guide". *Lancaster: Woodhead Publishing.*

[4] Stoller, D. (2004). "Stitch 'n Bitch: The Knitter's Handbook". *Workman Publishing Company.*

[5] Thomas M. (1938). Mary Thomas's Knitting Book. Dover Publications. New York.

[6] Zimmermann, E. (1972). "Knitting without Tears". *Simon and Schuster*, New York.

[7] Study on "Garment Sector to understand their requirement for Capacity building", http://texmin.nic.in.

[8] Gschwandtner, S. (2007). "KnitKnit: Profiles and Projects from Knitting's New Wave". *Stewart*, Tabori and Chang, New York.

[9] "Knitting". The Columbia Electronic Encyclopedia, Sixth Edition. (2003). Columbia University Press. Retrieved on 2006-12-21.

Textile Industries—Post COVID-19 and Textile Managements

B. Basu
Ex Scientist BTRA (Bombay Textile Research Association,
Ghatkopar, Mumbai–400086, India
E-mail: bibekananda.basu1502@gmail.com

ABSTRACT: *In India, all the industries including that of Textiles suffered a lot w.e.f. March'19 since the starting of the pandemic. The most sufferer were the unorganised sectors which accounts about 90% (depending upon the nature of the business) of the industries were almost came to the stage of loss making/closure. Their main reasons were the shortage of work force and stockpile up. Now at the stage of the ending the COVID-19, the garment sectors have started doing well, and most of them have reached at 100% working level especially the PPE manufacturers. The Organised sectors and the big houses are doing a good business in local and export market. The greater future is ahead of them. Those who could make proper planning and business strategy and manufacturing value added products are the best game players in the market today. The GDP of India was 1.6% (Q4, 20–21) and that of 20.1% in Q1 (21–22), 8.4% in Q2 and that of 6.3% in Q3, despite so many other constraints. It is expected to rise gradually as per IMF. Although Textile and Garment sectors have reached to certain level, the Core Industries like Coal, Natural Gas, refineries, steel, fertilisers, Cement, electricity etc. have shown Growth of 8.9% YoY June 20–21. The Textile sectors, both cotton, Polyester sectors, Machinery manufacturers are recovering from the shock but in comparison to 2019, it is still lagging in certain areas. The most of the Branded Organisations are doing very good to best business and they will grow more.*

There was a high impact on unemployment and uncertainty among the workforce in private sectors where some were not able to pay the salaries, some paid partially, but now in almost all cases, it is arriving at Normal Stage. The Education sectors are till now at high dissatisfaction stage and no prediction can be made right now (this is as per Dec'21 scenario). This paper has discussed all about the profitability, management's strategy to lead the industries at high level of satisfaction after COVID-19.

Keywords: COVID, Education Sectors, Future, GDP, Industrial Growth, Unemployment.

1. INTRODUCTION

The COVID pandemic had disrupted the normal life, industries, growth, employment, trade & business not only in India but across the world. The uncertainty prevailed at every stage of life. The factories especially the unorganized sectors closed the business, stocks piled up, the workers left the working places. The organized sectors slowed down the activities to certain extent, but those who were having good strategic planning, could run the organization by retaining the work force. The Personal Protective equipment (PPE) and medical textile manufacturers did zoom business.

The other way sufferers were front line workers like heath service, municipality staffs, policemen and state transport employees. The situational difficulties raised for the school & college going students and the "Online classes" started manufacturing half educated students! The online classes are lacking in quality education and remained just an eye wash!! However, the Pioneer Educational Institutions are firmer in running the Institutions by making compulsory presence of the faculties. As a result, several colleges started both online & offline classes for the benefit of the students.

Gradually with the slowdown of the pandemic situation, the migrant workers started to return, the organized sectors started running in full capacity & efficacy, but the unorganized sectors stated taking shape as per the situations. The GDP started to grow from deep negative to positive directions. Sales and Export potentiality started increasing in comparison to 2017–18, 2018–19.

The demand for the skill workforce and meritorious students especially for the Branded Colleges increased with better pay than earlier. The demand for the IT sectors, health services, supply chain, digital marketing are on ascending order. The work style is changing with "Work from home" saving the establishment costs. In this survey the above narrated gimps of COVID-19 social sock is explained in detail with present & future situations. [1-5]

2. THE IMPACT ON THE INDUSTRIES DURING COVID-19

During the lockdown period, there had been a migration of migrant workers going back to their respective places. Today almost 90% of them are back but and About 60% of the looms are working in Bhiwandi Power Loom Sector, and almost 90–95% looms are performing in Ichalkaranji, Malegaon and Solapur, Tirupur.

The Organised sectors producing daily use like Towels, Bed Sheets and PPE, Home Décor etc. were not much affected and now back to the near normalcy (as on 4th week of Nov'21). There is always good market for the Medical Textile globally.

The Polyester sector and Textile machinery manufacturing areas were affected, though not closed completely, but now they are at profit making stage (Q3 21–22). [1-5]

3. THE GROWTH RATE & GDP

The growth rate in 1st Quarter (Q1) 2021–22 was 20.1% followed by 8.4% in 2nd Quarter (Q2) and that of 6.3% in 3rd Quarter (Q3). The expected GDP in 4th Quarter Q4 would likely to be 6.1% with overall 9.5% GDP in the year 2021–22 as per the RBI Guidelines. It is due to the booming business of all the core and that of IT Sectors after the closures. The Export market has raised to the great extent at all sectors. [1-5]

4. THE NATURAL OF FUTURE WORK AFTER COVID-19: ACCORDING TO GARTNERIN TEXTILE INDUSTRIES

(By Modern Recruiters Team November 16, 2021)

There will be more of remote working especially for the white coloured jobs with more competitive executives with cut in compensation. The blue coloured people with competitive skill will on the floor. The employees in critical rolls would get more preference. Right man at right place with right compensation will increase the Organisation value. [1-5]

5. MIGRANT WORKERS & DECENTRALISED SECTORS

There are an estimated 139 million migrants in the country, according to the World Economic Forum. Indian migrant workers during the COVID-19 pandemic have faced multiple hardships. Millions of migrant workers had to deal with the loss of income, food shortages and uncertainty about their future during the complete lockdown. Most migrants in the country originate from Uttar Pradesh and Bihar, followed by Rajasthan and Madhya Pradesh. The cities of Mumbai and Delhi attract the highest number of migrants.

As new COVID-19 infections in the country decline, migrant workers have started returning in droves to the textile exporting hubs in Tamil Nadu and Uttar Pradesh, bringing cheer to the industry which is merry with seeing a 25–30% increase in export orders from the previous year (July 21). So many companies offered free travel tickets, housing, and food to run the organi-sations.

Apparel exporters said they expect the order position to reach the pre-pandemic level by the third quarter of FY22 (Economic times: Jul. 21, 2021).

As on today (30th December'21) almost all the organised and decentralised sectors are working at full capacity except that of entertainment and Travel Industries.

The unorganised sectors mainly consist of Agriculture, forestry, fishing will maintain its lively hood add to the economy. The construction, trade, tourism suffered but gradually recapturing. [1-5]

6. THE EMPLOYMENT OPPORTUNITIES AFTER COVID-19

The scopes of the employment will be higher after COVID-19 in the areas of Health Care, Social Workers, Tech & IT Jobs, Supply chain Management, Artificial Intelligence and Digital Marketing. The courses are available in all the Universities.

In comparison, the World's unemployment rate is 6.47% in the year 2020, that of 5.37% in 2019, 5.37% in 2018, 5.55% in 2017 which is not much different that of India. In India, the unemployment rate is 7.7% as 29th December 21 (Net search) which has increased after COVID-19. Now after running the industries at full efficacy, it is expected to go down gradually.

Mc Kinsey Global Institute says 15 "the pandemic accelerated existing trends in remote works, e-commerce, and automation, with up to 25% more workers than previously estimated potentially need to switch occupancy". [1-5]

7. THE INDUSTRIES ARE GROWING IN THE FINANCIAL YEAR 2021–22

Over 17,200 new companies set up in India during April–June 2021 (Press Trust of India, New Delhi Last Updated at July 19,21). More than 17,200 new companies were set up in the country during April to June this year while a total of 13.7 lakh companies were active at the end of June, according to official data. NIIT (National Institute of Information Technology) Limited announces Q1 (April – June 2021) Consolidated Results.

NIIT Limited—Consolidated Financials for Quarter ended June 30, 2021

(in Rs. Crore)	*Quarter Ended June 30, 2021*	*Quarter Ended June 30, 2020*	*YoY Change*
Net Revenue	301	201.8	49%
EBITDA	72.1	24.3	196%
EBITDA %	24%	12%	1188 bps
Profit After Tax	51.4	29.0	78%

EBITDA means Earnings before Interest, Taxes, Depreciation, and amortization (used as an indicator of the overall profitability of a business).

India's core sector growth up 8.9% YoY in June (business India today – 10th August 21). [1-5]

7.1 Textile industry pegged to grow 300% over next 2 years despite COVID impact, says report (Business India Today Monday, November 22, 2021).

The Business India Today report are given as follows:

7.1.1 After being hit hard by COVID-19, India's textile industry is well on course on the road to recovery. Indian technical textiles market could increase at a CAGR of 7.6% in Asia-Pacific to reach $23.3 billion in 2027, up from $14 billion in 2020, says the report.

Investment in value added services, e.g., marketing, warehouse rentals, logistics, courier, and other product fulfilment costs constitute a pre-requisite for the sector going to scale.

The textiles and apparel (T&A) industry contribute 2.3% to the country's GDP, 13% to industrial production and 12% to exports.

7.2 Some More Company News Showing Improvements

7.3 ***Net Profit of Vardhman Textiles*** rose 699.55% to Rs. 481.49 crore in the quarter ended September 2021 as against Rs. 60.22 crore during the previous quarter ended September 2020.

7.4 ***Reliance Industries*** Q2 Net Profit Jumps 46% to Rs. 15,479 Crore

The net profit, on a YoY basis, rose to Rs. 15,479 crore. IANS Published: 22 Oct. 2021, 10:11 PM IST.

7.5 09/08/2021. ***Indorama Ventures Plc (IVL),*** a global chemicals producer, saw its consolidated revenue grow by 52% year-on-year, and up 10% quarter-on-quarter, to US$3.56 billion.

7.6 ***Trident:*** Total Income – Sept 20 = 1171.44 and Sept 21 = 1662.57

7.7 ***Welspun India*** Q1 Net profit jumps over four-fold to 222.9 CR.

7.8 27/10/2021 · ***Raymond*** Q2 profit at Rs. 56 crore total income was up over two folds to Rs. 1,583.26 crore. Raymond Ltd. on Wednesday reported a consolidated net profit of Rs. 56.15 crore for the second quarter end.

7.9 ***Lakshmi Machine:*** Standalone September 2021 Net Sales at Rs. 792.42 crore, up 108.92% Y-o-Y. 26.07.2021. Lakshmi Machine Consolidated June 2021 Net Sales at Rs. 457.62 crore, up 171.14% Y-o-Y, 29.10.21.

7.10 ***The Arvind Mill's*** revenue from operations during July–September 2021 stood at Rs. 2,115.14 crore, a jump of 62.05% as against Rs. 1,305.17 crore in the year-ago period.

7.11 TCS Net Profit Jumps 29% to Rs. 9,624 Crore in Second Quarter

TCS net profit surged 28.75% in its second quarter (Q2) profit—ended on September 30—of the current financial year 2021–22 (FY22).

7.12 ITC Net Profit Rises 14% to Rs. 3,697 Crore in September Quarter, Revenue Up 11%

ITC Limited Q2 Results: The cigarette segment reported a 10.2% growth in revenue at Rs. 5,642 crore compared to Rs. 5,121 crore in the corresponding quarter last year.

It is observed that after the set back in the year 2019–20 and that of 2020–21, there is booming period in the Q2 of 2021–22 for almost all the industries.

7.13 The Times of India dated 2nd January 22 has shown that the ***Car sales*** have increased from the year 2019 (Start of the pandemic) and 2020 (full year of pandemic) in the year 2021 though lesser than that of 2017 and 2018.

All these rises in sales, turnover and profit margins are in progress because of the opening of the market, trades, IT Sectors after the pandemic situations. The vacuum gaps are going to be fulfilled.

A survey was conducted by the author in 24 Spinning and composite mills in North India revealed that in comparison to 2019–20 to that of 2020–21, almost all the companies has shown less profit. The Nitin Spinner, sportking, Nahar Spg. Maral overseas have made profit. It was because of the better quality, product diversification, Marketing Policy and Managerial capacity. The Sick mills are working but to come at level. The year to year 2021–22 has shown good profit to almost all companies which will be declared after March 22.

The latest Market survey reveals that Vardhman is putting another 65510 spindles, Trident Budhni is putting 1 lakh spindles, RSWM is expanding with 72,000 spindles, Sangam is putting 30,000 spindles, Nahar Spinning is further adding 6000 spindles. All are doing very good. As a result, the LMW, Reiter are also doing well (*Source:* www.Moneycontrol.com).

8. ITF TIMES

Shows in Q1, 21, the Export has gone up by 42% comparison to Cotton yarn, fabrics, Garments (Made up), H.L. Products (High quality fabrics, Industrial Fabrics, Technical Textile, Speciality fabrics etc.) in comparison to Q1, 19 and that of Q1, 18. Similarly % of Exports has also increased by 18% in Q1,

21, comparison to Q1, 19 and that of 18 too in case of Manmade Yarns, Fabrics and Made ups.

All the prospects are getting possible due to the increasing market demands, Management's new policy decisions, Mission, Vision, and strategic Planning. Most of the management's policy is to go for new Product developments, Modernisations, Skill work force, Strong R&D, and recruitment policies. Those who are lagging in those aspects are non-profit making organisations. However, the health care Industries always remain at Profitable stage including that of Pharmaceuticals. [1-5]

9. THE JOB OPPORTUNITIES OF THE PROFESSIONALS TILL JANUARY 2022

It is found that since Sept.–Oct. '21 to till now there is a favourable time for the Engineers in the IT as well as in core sectors. The Job opportunities were very less since pandemic was started i.e., June/July 19 and that of the whole year of 2020 because of the slack in the business and trades. Today apart from IIT/IIM/NIT/NIFT etc., the other engineering colleges can send their fresher's in the industries with higher Pay than that of previous years.

One thing is noted that the MNC and that of IT sectors are in search of talents who are paying better remuneration to the deserving candidates.

9.1 The Times of India published that "It is the best placement Season ever at IIT. The reason is the same that the Organisations need the talents to run their Industries in much Professional ways than earlier because of the sufferings.

9.2 One reputed Engineering College in Maharashtra has proudly declared their Placements of 2021 & 2022 batches* with higher pays shown here.

Total offers received = 800 + during August 20 to Dec. 21

Placement offers to 2022 batch till 22/12 = 400 +, Highest package = 20 LPA

No. of companies who offered placement = 45, Number of drives going on = 10+

Mission 400 achieved, *Total offers of 2021 batch = 407+*

Their Recruiters*:

Cognizant, TCS, Persistent, Capegemini, Infosys, Wipro, Goldman Sach, Dassult systems, TIAA, Reliance, Welspun, RS. Lab, NeML, Hexaware, Iauro, Burns Macdowell +++++

This new tendency has generated very recently. (1) There are other side also that some students of avg. merit are ready to join in lesser pay because of the poor job opportunities in certain places. The female students are mostly affected. (2) Most of the students are showing interest in Mktg, PPC, Merchandising, etc. rather than to join in Production and shop floor.

On conclusion it can be said that in the coming years there will be huge job opportunities with handsome pays as the industries are going for more and more innovative ideas, product developments for the higher profitability and to cope up with this, they prefer Talented students. [1-5]

10. ROAD AHEAD

India is working on major initiatives, to boost its technical textile industry. Owing to the pandemic, the demand for technical textiles in the form of PPE suits and equipment is on rise. Government is supporting the sector through funding and machinery sponsoring. Similarly, the Medical, Industrial Textiles and that of Smart Textiles are in high demand.

Top players in the sector are attaining sustainability in their products by manufacturing textiles that use natural recyclable materials.

The future for the Indian textiles industry looks promising, buoyed by strong domestic consumption as well as export demand. With consumerism and disposable income on the rise, the retail sector has experienced a rapid growth in the past decade with the entry of several international players like Marks & Spencer, Guess and Next into the Indian market.

High economic growth has resulted in higher disposable income. This has led to rise in demand for products creating a huge domestic market. [1-5]

11. THE UNCERTAINNESS PREVAILS

COVID 3rd wave may pull GDP growth down to 7%: Economist Abhijit Banerjee

Press Trust of India | Kolkata Last Updated on August 5, 2021, 22:35 IST27

Nobel laureate economist Abhijit Vinayak Banerjee apprehended that the impending third wave of the COVID-19 pandemic might adversely impact the GDP, and its growth rate might go down to 7%, even below the IMF's recent projection of 9.5%. The novel coronavirus has a new name and new features now. Omicron is a highly transmissible mutated version of the COVID-19 virus.

After almost two years of restrictions, the world was beginning to relax. Business opportunities arise. People around the world are getting vaccinated at a record pace. In India, more than 10 million people have been vaccinated. The economy had picked up and things were about to get better, when the new COVID-19 mutant, Omicron destroyed the hope. (As published by Dinesh Exports 28)

The Textile Industry is heavily dependent on MSMEs. They play a major role in producing items for domestic and international markets. At a time when the global supply chain is already impacted with COVID-19 and lockdowns,

Omicron is a threat to the industry. The second wave in the month of April–June '20 had already crippled many businesses and workers. With great efforts and help from the government, it had started to pick up. Now the Omicron is expected to disrupt the industry again.

Let us hope for the better (This paper was written from November 21 to 1st week of January '22).

As on December last week '21, the unemployment in India was 7.9% which was 6.3% in 2018–19 and 4.7% in 2017–18 (CMIE data). The world figure of unemployment found to be 6.5% in the year 2021 and ILO predicts the same to come down to 5.7% in the year 2022. [1-5]

12. CONCLUSIONS

COVID-19 created a setback for the industries, educational Institutions, and in human life. Although the Unemployment % shoot up, migrant workers suffered a lot, but the normalcy has almost arrived gradually. The industries have taken its own shape some of them have started making larger profits than the previous golden years before Carona. New and newer industries are coming up, some are expanding. The Export potentially started growing. The Education sectors were coming to near the normal but again it got disrupted (January 1st week'22). The Scopes of the bright students started increasing who passed recently with new strategic planning of the Industries in both IT & that of Core sectors.

Now style of working of the industries will differ in comparison to 2018/19 that work from home will be at higher stage for the white-collar jobs. The more scopes are expected for health-related workforce, Digital & media, IT, Supply chain Managers, and Artificial Intelligent Officers.

The 3rd wave, Micron has started the path of disruption in education, hotel and tourism sectors and no predictions can be made right now. The economists are forecasting again the downward tendency of GDP. Let us hope for the betterments.

REFERENCES

[1] RBI Guidelines on GDP, Handbook of statistics on Indian economy, 2021, https://www.rbi.org.in/scripts/AnnualPublications.aspx?head=Handbook%20of%20Statistics%20on%20Indian%20Economy.

[2] Economic Times, 21st July 2021.

[3] Press Trust of India, New Delhi, 19th July, 21.

[4] Business India today 10th Aug. 21.

[5] Times of India, 22nd January 2022.

[6] Press Trust of India, Kolkata 5th August 2021.

Promotion of Energy Conservation Technologies

Bhavdeep Shah
Man-Made Textiles Research Association, Ring Road, Surat, India
E-mail: director@mantrasurat.org

ABSTRACT: *The importance of energy conservation awareness at home is relatively easy to understand, but its thorough implementation is usually delayed in the early stages of energy conservation awareness programs in factory. Therefore, for concrete implementation, it is desirable to develop company-wide coordination measures similar to QC activities in factories. In addition, in order to promote energy-saving measures efficiently, it is effective to separately consider a general management method for "reasonable use of energy" and a process-specific method to be developed for each specialized field.*

Keywords: Energy Conservation, Textile, Automation, Heat Exchanger, Steam.

1. INTRODUCTION

Energy conservation is a significant step to overcoming the vital problem of the worldwide energy crisis. In particular, Sub continent countries i.e. India are interested to increase their awareness on the inefficient power generation and energy usage in their country. However, Data of energy use are limited for proper execution of energy conservation.

In developing countries energy saving and conservation technologies should be spread to all governments' organization as well as industrial owners and technocrats. It is essential that they gain practical knowledge of the presently available energy conservation technologies.

Due to high energy intensive industry i.e. textile sector, it is recommended to implements energy conservation technologies using modern energy saving techniques.

For example:

- Regulating the temperature in the steam pipes,
- Adjusting the air/fuel ratio in the boilers, and
- Installing heat exchangers using warm waste water.

There are several ways to conserve or saved energy in textile sector out of which few are discussed below.

2. ENERGY CONSERVATION MANAGEMENT TECHNOLOGIES

2.1 Organizational Validation

The energy management is applicable to a various departments of the company, it is necessary to improve the awareness and knowledge to obtain the participation and co operation of everybody involve in the operation. So it is essential for technical person with expertises with technical knowledge to lead the role of energy conservation, the implementation of an energy conservation program should be company-wide, for example organizing of an 'Energy Management Committee'.

2.2 Improving Efficiency in Industrial Energy

2.2.1 *Lighting*

The introduction of LEDs in the industrial lighting has achieved significant savings in electricity in the industry. With recent research and development electricity saving LEDs can save quite a few percent for the existing level of illumination. In general, the efficiency of illumination is affected by a range of parameter, such as light source intensity, coefficient of the reflection, shape of the lamp, room layout, interior finish, colour and the distance of light source. Thus it is extremely important to re check all the parameter related to the efficiency of the light source to save the electricity such as reducing the number of lamps in use by switching from global lighting to local lighting as much as possible.

2.2.2 *Electric Motor*

Electric motors used in the textile industry are relatively smaller; though conventional machine had driven by a single motor with the help of generated mechanical power to give power to different parts of the machine in sequential manner. Latest machines uses multiple motors with distribution control system, that is directly linked to machine drive independently which is further synchronise with machine parts to drive independently from others. This is also a vital feature in energy conservation. Selection of each motor mainly depending up on mechanical performance and proper motor capacity. This leaves significant scope for a review in energy conservation.

2.2.3 *Electrical Heating*

In the textile industry, electric heating has been replaced by other methods for some time in order to achieve cost reductions for example steam, gas heating, or direct or indirect fired heating in most of the process. However, being less complex and initial cost of electrical heating machine is low compare to other process; therefore it is popularly used in small heating applications. [1] Therefore, it is desirable to explore into unconventional heating methods, such as far-infrared radiation heating, high frequency dielectric heating and microwave heating.

2.3 Energy Efficiency in fuel

2.3.1 *Fuel Selection*

Utilization of fuel in textile industry switches towards oil from coal. For efficient energy investigation for the revival of coal for further move to liquefied and other city gases due to variation of fuel prices. Selection of fuel in such a way that have potential to improve the characteristics of fuel high calorific value and ease of combustion to minimize air pollution. [2]

2.3.2 *Boiler Selection*

Textile industry has changed system to use forced circulation instead of natural circulation boiler. Because of this efficiency has improved quite significantly over the time. Although In high performance boilers scaling rate inside the water tubes is quite high therefore feed water management is necessary. Moreover this boiler has high evaporation ratio so that it can be fully automated including feed water and combustion management.

2.4 Energy Efficiency in Steam

2.4.1 *Piping*

The textile industry uses steam but not in larger quantity though steam losses due to heat energy from steam moving pipes and pressure drops are considerable.

Therefore, it is required to use high pressure and minute diameter rather than low pressure and large-diameter piping for steam transportation over long distances with pressure dropping valves placed as necessary point of distance to regulate the steam pressure for reduction heat losses. Also, as pressure losses around bends are great, it is advised to make their bends large.

Due to thermal expansion Steam joints should be placed in such a way that steam leaks could be avoided. Additionally, use of correct heat insulation material for efficient use is necessary to maintain the temperature inside the valve tank and treatment tank as well as the piping

2.4.2 *Steam Accumulators*

Since steam is often used in dyeing work, the amount of steam consumed fluctuates greatly during the work time. Highly efficient water tube boilers are designed so that almost no water remains inside the boiler but the boiler cannot act in response to momentary and sudden load changes, though responding to automatically controlled slow load changes is not a problem. In such cases, a steam accumulator can be installed in between the boiler and the heat-consuming load to convert excess steam into hot water and store it when the load is light. This again transforms the heated water back to steam when the load is heavy in order to strengthen supply to the load. This allows the boiler to operate continuously at an average load, which is extremely beneficial in terms of energy saving.

2.4.3 *Drain Recycling*

Steam has been discharged after consuming heat energy. However, from the viewpoint of energy saving, it is necessary to collect and reuse the heat energy carried by the wastewater. [3]

2.5 Heat Exchanger Utilization

In each manufacturing process in the textile industry, frequent heating and cooling of gases and liquids as heat carriers is required. It does this by exchanging heat between different fluids and uses heat exchangers to perform indirect heating and cooling to avoid contamination or chemical reactions due to their direct contact. [4] It is important to use the right heat exchanger for the right purpose.

2.6 Measuring Instruments and Automatic Control

Energy saving is an activity that allows to accurately capture and quantify actual energy consumption in a plant and take improvement actions to rationalize and save that energy. While measuring instruments are necessary to obtain quantitative data, the study of the use of sophisticated measuring instruments draws on recent developments in mechanical and electronic engineering, combined with automated control systems will become ever more important. [5]

3. CONCLUSIONS

The transfer of these technologies could be achieved through:

1. Conducting surveys of energy usage and efficiency at the plant level;
2. Preparing manuals on energy management and energy conservation/ saving technologies, based on the findings of the above survey;
3. Presenting and discussing the manuals at seminars held for government officials, representatives of industries, plant managers and engineers;
4. To disseminate the manuals to other developing countries for their proper utilization and application.

REFERENCES

[1] Energy audit team, Energy Audit Study of Man-Made Textiles Research Association, Surat, 2019.
[2] Goldstein, H.B. and Smith, H.W. (1980). "Spray System: TCC", 12–49.
[3] Dempsey, E.P. and Vellings C.E. (1975). "Heat Transfer Printing", *Inter print*, 46.
[4] Energy Use, Loss and Opportunities Analysis US DOE, 2004.
[5] CII Energy Efficiency Bulletin.

Sustainability, Knits and Fashion—An Approach towards Sustainable Development in Knitting Industry and Fashion

Sayantika Debnath and Sankar Roy Maulik*

Department of Silpa-Sadana, Visva-Bharati (A Central University), West Bengal, India

E-mail: *s_r_moulik@yahoo.co.in

ABSTRACT: *The textile and apparel industry of the whole world is dominated to some extent by India, making it one of the largest producers in this sector. The domestic and apparel sector contributes about 5% to India's GDP, making this industry one of the largest employers in the country. The knitted textile industry is an ever-growing industry in India, and it comes in the second position after the woven counterpart. Gradually knitted fabric and garments are gaining popularity because of several advantages that it has over woven textiles. Consumers are more drawn to sustainable fashion, which is safe for them and the environment, rather than cheap synthetic garments. Natural fibres, dyes, and finishing agents are some of the key ingredients to achieve the goals of a sustainable and green industry. Fluidity has been taken as an interpretation in this work. The inspiration is taken from a wide variety of situations, landscapes, textures, and patterns in nature that are in constant change. 'Liquid lines' is a collection of women's wear that draws inspiration from the word 'Fluid' and its interpretation in various fields. It is an exciting and edgeless amalgamation of natural colours, textures, and techniques that imparts a lot of visual appeal and character. Developments in fashion and manufacturing technology can go hand in hand, making the whole process enjoyable and satisfactory for the environment, manufacturer, and the consumer as a whole, where there is less waste and harm to tackle at each step. Hence the textile industry needs to transition into an eco-friendly environment through small changes to better our earth.*

Keywords: Cotton, Dyeing, Fashion, Knitting, Printing. Sustainability.

1. INTRODUCTION

The present scenario shows that an approximately 1% share of textiles is only being dyed with natural dyes that are used mostly in the cottage sector by traditional artisans and small-scale textile dyers. Reasons for not using natural dyes in the large-scale textile sectors are unavailability of readily extracted and purified dye powder, lack of standardized dyeing processes with the assurance of reproducible and uniform shades and non-warranty of the required level of colour fastness to achieve. [1] However, consumers nowadays are becoming more and more conscious about the products they are using and leaning towards organic, sustainable and eco-friendly products. A sustainable textile product can be defined as one that is created, produced, transported, used and disposed of with due consideration to environmental impacts, social aspects and economic implications. Thereby it will satisfy all three pillars of sustainability, and is expected to create the minimum possible or very least environmental and social impacts throughout its entire life cycle. To make this life cycle sustainable for any textile goods, all these inputs and outputs are to be environment-friendly and eco-safe particularly to the living bodies on earth including humans for the entire life cycle. Therefore, it should be non-carcinogenic and has to utilize renewable/recycled input streams with the least energy consumption. The knitting industry has emerged as a premier supplier of textile value chains and it occupies an important position in the global textile industry. It also makes a significant contribution to the economy and employment of our country. [2] In general, most of the natural dyes used for dyeing textiles are obtained from vegetable origin. Judicious use of natural sources in dyeing and printing leads the process to be organic and hazard-free. [3] In view of the above, the main objective of the article is to use these organic colours for dyeing and printing knitted fabric through an environmentally friendly route.

2. MATERIALS AND METHODS

2.1 Materials

2.1.1 *Fabric*

100% scoured and bleached single jersey knitted fabric, cotton French terry knitted fabric and non-bleached cotton French terry knitted fabric was used in this project. These were obtained from the local market and through an online shop from Delhi, India.

2.1.2 *Natural Dyes*

The natural dyes were sourced from www.amazon.in and also bought from the local market.

Table 1: Details of the Plant Used for the Experiment

Species (Family Name)	*Common Name*	*Part Used*	*Colour Obtained*
Acacia catechu	khair, cutch wood	heart wood	Reddish brown
Allium cepa	onion	onion bulb scales	Pinkish orange
Caesalpinia sappan	sappan wood	wood	Red
Bixa orellana	annatto	seeds	Peach/Orange
Camellia sinensis	tea	leaves	Light brown

Fig. 1: Pictorial Representation of (1) *Acacia catechu*, (2) *Allium cepa*, (3) Caesalpiniasappan, (4) Bixa orellana, (5) Camellia sinensis

2.1.3 *Mordants*

Inorganic salts viz., ferrous sulphate, copper sulphate and aluminium sulphate, and bio mordanting agents like myrobalan and pomegranate rind were used as mordanting agents during this study.

2.2 Methods

2.2.1 *Aqueous Extraction of Dyes*

After sourcing the natural dyes, the extraction of the colouring compound was performed separately for individual dyes. 100 g of dyestuff in case of crude source and 5 g of powdered dye were taken for extraction purposes. Those specified quantities of vegetable sources were added to 1 litre of water and the solution was boiled at 95°C to 100°C for about 45 min and then cooled to some extent. This solution was later filtered through a nylon mesh and the solution was used for preparing the dyeing bath.

2.2.2 *Dyeing Process*

In case of metal based mordants, post mordanting method was performed, and in case of bio mordants, pre mordanting was done. Mordanting process was carried out for 20 min both in hot and cold conditions. Dyeing of knitted fabric was done for one hour at 90°C keeping a fabric to liquor ratio of 1:30. After dyeing and mordanting, the dyed samples were washed thoroughly with cold water followed by a hot wash with non-ionic detergents and then cold washed again.

3. DESIGN CONCEPT

The term 'Fluid' is the key inspiration for the collection of women's wear that was prepared for this project. This collection aims to interpret the fluidity found in nature. The ever-changing landscapes, textures combined with the absolutely beautiful natural colours give the collection lot of aesthetic appeal. Fluidity in nature is best described with its ability to change. Change is indeed constant and in human lives. This collection also tries to incorporate silhouettes that give the female body a sense of independence, appeal and freedom.

3.1 Design Process

The design process involved a variety of steps which included the selection of the fabric, sampling with various natural dyes and mordants, brainstorming on the theme and mood of the collection of garments that are to be designed, illustrating the garments, finalizing them for pattern development and cutting and ultimately stitching of the garment.

3.2 Brainstorming

A brainstorming session is a great way to generate new ideas, modify the old ideas and finalize the product range. One can get a clear vision of the work through this method and come up with the pros and cons of design which helps the designer to rethink their ideas. In this project brainstorming about the concept of design helped to a great extent for choosing the techniques, textures and theme to base the work on.

Fig. 2: Shades of Natural Dyes Developed on Cotton Knits

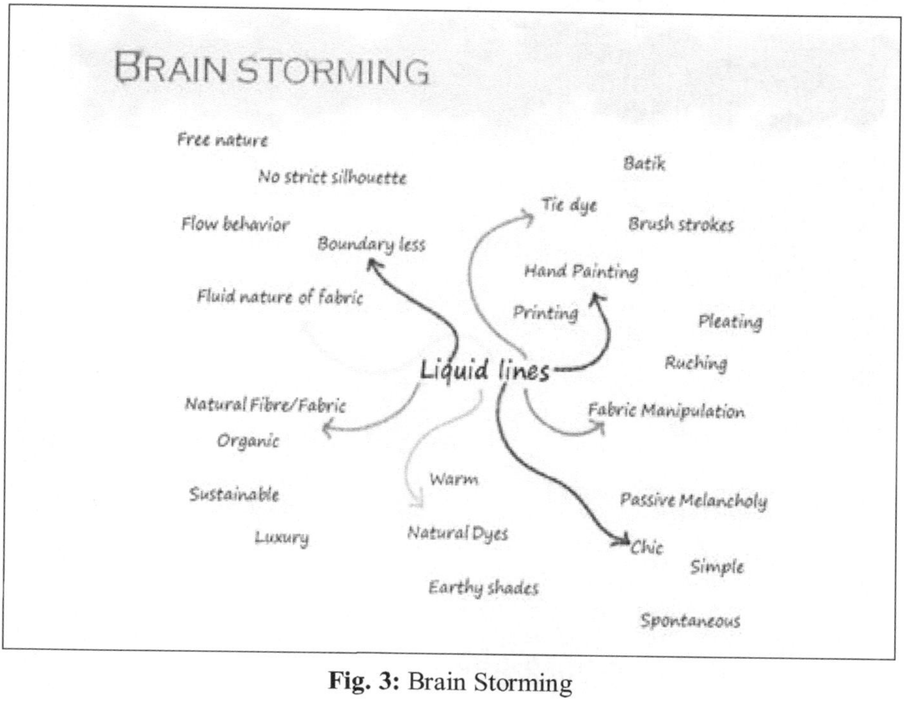

Fig. 3: Brain Storming

3.3 Inspiration Board

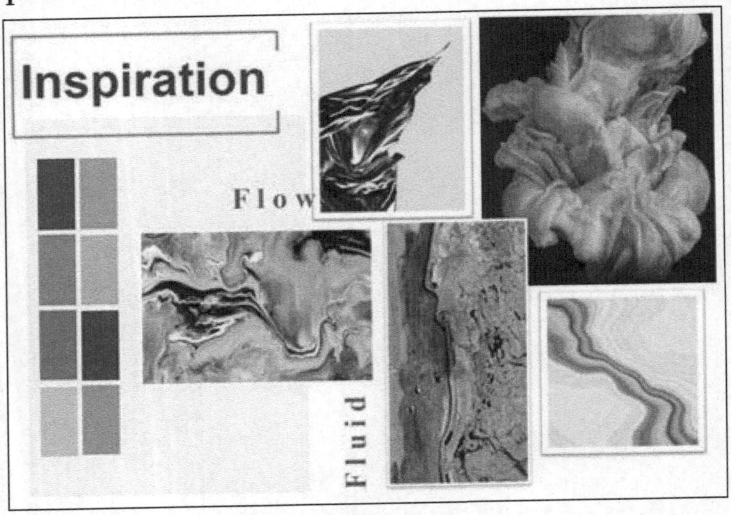

Fig. 4: Inspiration Board

3.4 Illustration of the Products

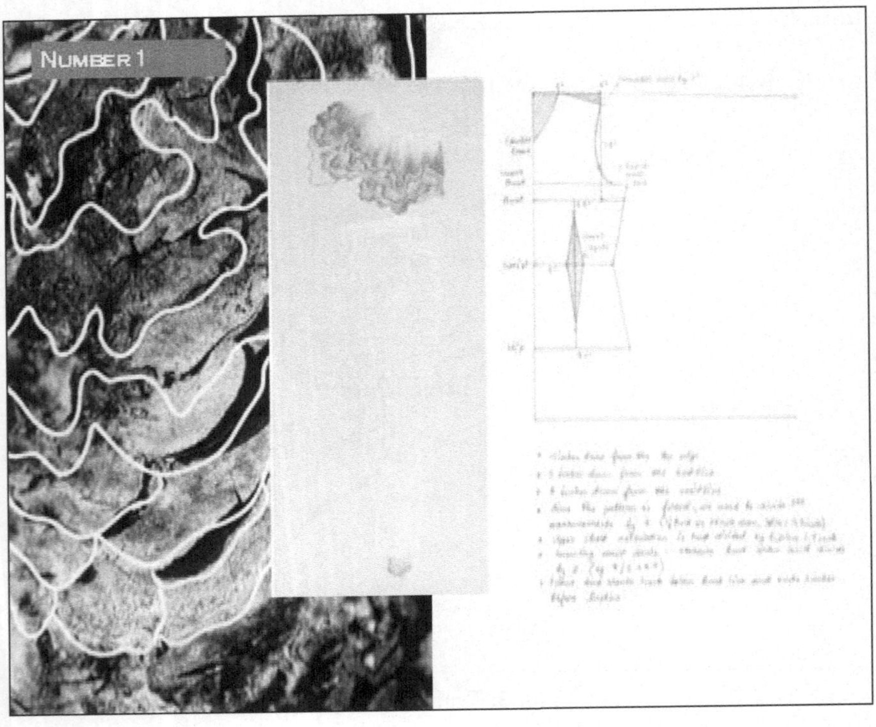

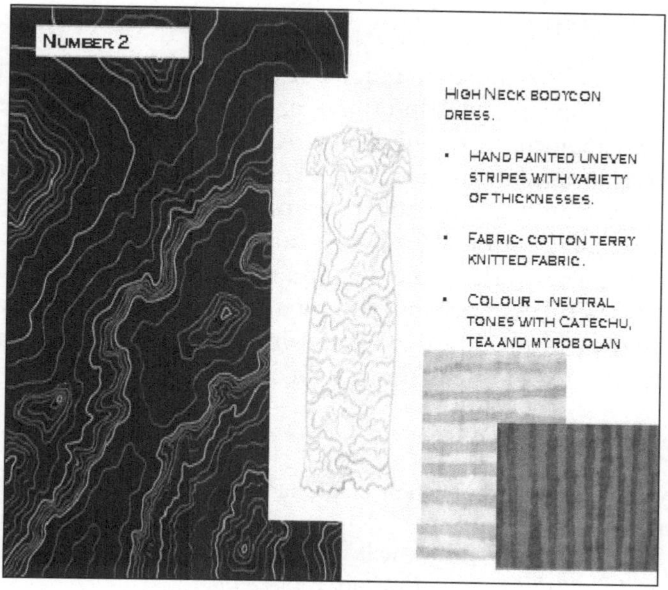

Fig. 5: Illustration of the Products

A graphical representation is provided in Figure 5 for case of understanding.

3.5 Designing of the Garment

Dimension of the garment and pattern cutting is an important step in apparel designing. First the colour of the collection was determined with the help of the developed samples and then the patterns were developed for the fabric to be cut and sewn.

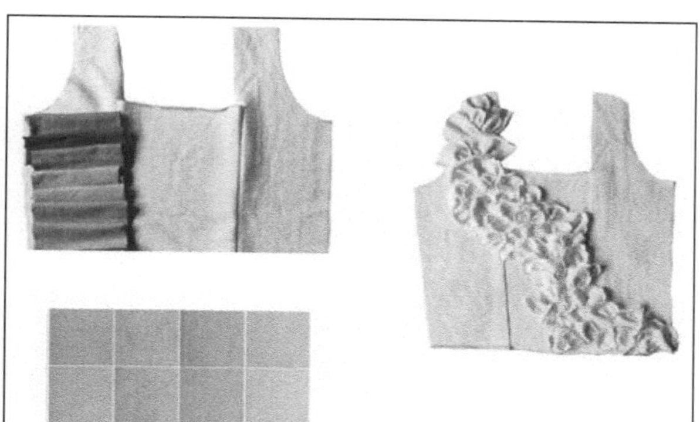

Fig. 6: Designing of the Garment

4. CONCLUSION

This project not only explores the scope of natural dyes and natural mordants on cotton knitted fabrics but also utilizes an extensive design process in each step of the development of the garments. As per prior researches, the natural dyes and mordants used in this project are known to possess good antimicrobial properties. This comes as an extra layer of protection in the present world which is struggling with COVID-19 along with many other harmful microorganisms. The world is moving at a rapid pace where consumers seek sustainability, safety, comfort and fashion all at once. These projects are just another step towards making the apparel industry green and sustainable for future generations to provide them with an eco-friendly fashion.

REFERENCES

[1] Roy, M.S. (2015). "Application and Characterization of Natural Dye", *Asian Dyer*, 12(3), 30–37.

[2] Ray, S.C. (2012). Fundamentals and Advances in Knitting Technology, *Woodhead Publishing India, Pvt. Ltd, New Delhi.*

[3] Prabhu, K.H. and Bhute A.S. (2012). "Plant-Based Natural Dyes and Mordants: A Review". *Journal of Natural Production Plant Resource*, 2(6), 649–664.

Epilogue

This section of the book compiles the work of various researchers and their finding. The book of papers contain the papers related to the progress of work in the field of medical and apparel textile. A wide variety of papers are selected looking to the need of the day in this post COVID era.

1. In this paper researchers describe about the Cornhusk fibre with their unique properties which was discarded as waste. The eco-friendly and bio-degradable fibres were extracted successfully from the cornhusk using NaOH at different concentrations. The optimized results were obtained at 1.25% NaOH concentration which shows satisfactory values of Tex and diameter with optimum yield percentage. Tensile strength and elongation are also better. Structural analysis confirms the presence of important chemical groups of lignocelluloses fibres and a crystalline region respectively.
2. In this work HDPE, plasticizer, antioxidant was used to improve the spin ability and draw ability of HDPE polymer by sustainable alternative approach of melt spinning to overcome the drawbacks of so-called gel spinning of virgin HDPE. The stearic acid interferes in this polymer chips to provide lubrication properties of HDPE/SA blend which result increase in spin ability and draw ability of blend filament. Stearic acid has zero toxicity and user freely.
3. The author reported on phase change material coated nonwoven interlined material for the study that shows the good potential of heat absorption for the intrinsic change in the temperature near 30 degree Celsius. The given fabric shows the significant increase in the stiffness which limits its application but as a interline material in the form of the panel there is a scope for its application.
4. In his study, the author showcases the uses of low cost scoured cotton to be used as sanitary napkin in terms of their water retention efficiency (WRE) and fibrous absorbing part sanitary napkin type 1 (FSN1 is compared to virgin cotton (VC) and fibrous absorbing part infused with SAP granules of sanitary napkin type 2 (FSSN2). It has been observed that WRE% increased 3.6 times from VC to SC by simple scouring. SC can compete with FSN1 as both of them provide 720 ± 31.62% and 828 ± 30.33% WRE respectively. This preliminary study indicates SC low cost, biodegradable absorbing material for the manufacturing of low cost sanitary napkins.

5. The author reported the need of green energy using Textile with Polymer Electrolytes for wearable batteries. The work stated the preparation of the flexible C60 dispersed polymer salt and the maximum ionic conductivity was obtained for 20 wt% of C60. It was analysed that the flexible nature of the polymer electrolytes films with a good approximation of conductivity can be used for various e-textile applications.
6. In this study, fiber fineness of 'okra' bast fiber was measured at three different test lengths (1.5 mm, 10 mm and 100 ± 20 mm). Variation in the fiber fineness was observed with the test lengths. Fineness of the okra fiber was found proportionate to the test lengths.
7. This paper reported on effect of dyeing on properties of polyester/viscose blended yarn and it has been observed that the yarn properties except yarn unevenness and imperfections significantly change after dyeing, irrespective of depth of shade. Yarn spinning technology is found to have most influence followed by blend proportion and depth of shade.
8. PPy coated jute/SD sorbents are prepared and found as effective filtration material. This paper focus on removal of heavy metals from waste water. PPy coated jute and saw dust shows 97% and 99% Cr removal efficiency respectively.
9. In this study aramid yarns which are difficult to dye, is pre-treated with solvent Dimethyl formamide to facilitate the dyeing process and then dyeing is done with cationic dye. Pre-treatment and dyeing were done at different temperatures and time to see how these parameters affect the process. The fastness properties of dyed fibers found better at high dyeing temperature and time followed by pre-treatment temperature and time.
10. This paper review various technologies and approaches used in processing of textile fabrics to achieve sustainable production. The paper reported about green approaches to textile processing i.e. use of green fibres, chemicals, solvents and optimization of the processes, adopting reuse and recycling of process waters. It also includes use of green fuels and its combination of all above mentioned approaches. It also discusses latest trends adopted by industry.
11. This research work is directed to the use of dye extracted from Indian madder root powder and found to have acceptable dyeing properties producing excellent shades of red and black colour. Significant difference in colour strength of dyed woolen fabric samples as a result of application of different mordants were observed. It has reported that alum mordanted samples show higher colour strength as compared to ferrous sulfate mordanted and unmordanted woolen fabrics.

12. In this review paper author reported on woven and knitted structure and preparation compound fabrics. The author claimed the advantages that by compounding may give better result and this compounding can be done in various ways like lengthwise, width wise and layer wise. The special study can be carried out on the joining of these structures.

13. In this review paper authors presented a short documentation about the contemporary staple fiber spinning system. It has highlighted various spinning methods, their limitations, yielded yarn structures and their properties.

14. This research paper reported on the manufacturing and use of new jute technical textile products using braiding technology. This paper emphasizes on the development of a particular product, named, *jute braided sapling bag* which is a tri-axial braided construction to develop *nursery sapling bag* from jute. This product is meant as an eco-friendly substitute of plastic-based nursery sapling polybags which are commonly used in agri-horticultural.

15. This review paper provided a discussion on various methods of self cleaning finishing on the textile fabric and their application on various products. It has also narrated about green option which can help to save money on dry cleaning and laundry expenses, as well as energy and time.

16. This study explains the application of nano formed ceramic material on the variety of jute fabrics to improve the performance. A novel aqueous based titania application method was developed to impart hydrophobic property. It is observed that there is significant improvement in UV protective property. Hydrophilicity studies showed high contact angle and Titania finish did not deteriorated the tensile strength of fabrics.

17. This paper emphasises on surgical gowns for healthcare and medical applications. The majority of gowns are now made of polyester or a polyester-cotton blend. Fluid-repellent chemicals are applied to cotton muslin and tight weaves have traditionally been used to make woven reusable surgical gowns. Product design difficulties relating to comfort are especially concerning for protective clothing, as comfort is a fundamental need of the product.

18. This paper reviews the procedure of manufacture and their availability in India application of antimicrobial finish on textile. This paper provides the reader with a detailed understanding on the role of nanotechnology and its standing in the present magnitude of COVID Virus, the types of applications and incorporation of nanotechnology in the antimicrobial textiles along with the treatments, chemicals, and substrates used.

19. The review paper provided readers a small summary about opportunities and challenges in medical textiles in India. One of the emerging and fastest-growing segments in the field of technical textiles is medical textiles. With the outbreak of COVID, medical textiles had become a national and international emergency. The study emphasized on the business opportunities in medical textiles.
20. This paper reviews on smart textile that sense if there is any change and at the same time react to those environmental conditions or stimuli, such as mechanical stress or strain, thermal, chemical, electrical resistance or capacitance, magnetic or other sources. It reported that there is huge potential application of smart textile in the area of smart PPE kit to prevent spreading of virus by eliminate contact while monitoring psychological parameters.
21. This research focuses on the development of sportswear materials with enhanced antibacterial and aromatic properties. Four type of fabrics were selected for the study and fabrics were finished with antibacterial agents and aroma tea tree oil. It is observed that cotton single jersey knit fabric is more suitable for performance sports fabric.
22. In this review paper the different types of reinforcing agents have been used to produce hybrid textile reinforced composites. The production, types of hybrid composites, the hybridization effect, performance properties, and diversified applications of hybrid textile reinforced composites are reviewed in this paper.
23. This work explains the effect of developing flexible electrically-conductive non-metallic textile fabric by two different processes. It is noted that lowest resistivity is achieved in case of electrochemical polymerization in spite of higher surface roughness. There is a change in resistivity observed as the *in situ* electrochemical polymerization process modified with graphene.
24. Natural dyes are emerging globally because of the growing consumer awareness about environmental conservation and control of pollution. Efforts should be made to identify the raw materials from plant sources and to standardize the recipe and dyeing conditions for their use. In this research paper, it is reported that painting with natural dyes contributes to value-added textiles.
25. The study of this review paper describes the seamless technique as revolutionary technique because of its smoothness, comfortable fitting, invisibility and easy care properties. The world wide trends predict that seamless garments will be considered as most popular garments especially

among the youth. The seamless garments have potential to cover fifty percentages of the industry's sales in coming ten years by exploring more technical trainings with value added facilities.

26. Review paper deals with the scenario of textile industries in Post COVID-19 and described the suffering of the unorganized sector due to shortage of work force. However the garment sector is improving in the post COVID era. The Textile sectors, both cotton, Polyester sectors, machinery manufacturers are recovering from the shock but in comparison to 2019, it is still lagging in certain areas. The most of the Branded Organizations are doing very good to best business and they will grow more.
27. This paper describes the importance of energy conservation and producers for implantation. It suggested companies to have similar types of QC activities in factories. In addition, in order to promote energy-saving measures efficiently, it is effective to separately consider a general management method for "reasonable use of energy" and a process-specific method to be developed for each specialized field.
28. This project not only explores the scope of natural dyes and natural mordants on cotton knitted fabrics but also utilizes an extensive design process in each step of the development of the garments. This gives as an extra layer of protection in the present world which is struggling with COVID-19 along with many other harmful microorganisms. These projects are just another step towards making the apparel industry green and sustainable for future generations to provide them with an eco-friendly fashion.

Author Index

9 789390 951390

Printed at Repro India Ltd.